WI-FI ~TM~, BLUETOOTH ~TM~, ZIGBEE ~TM~ AND WIMAX ~TM~

WI-FI$_{TM}$, BLUETOOTH$_{TM}$, ZIGBEE$_{TM}$ AND WIMAX$_{TM}$

by

H. LABIOD
ENST, Paris, France

H. AFIFI
INT, Evry, France

C. DE SANTIS
INT, Evry, France

A C.I.P. Catalogue record for this book is available from the Library of Congress.

ISBN 978-1-4020-5396-2 (HB)
ISBN 978-1-4020-5397-9 (e-book)

Published by Springer,
P.O. Box 17, 3300 AA Dordrecht, The Netherlands.

www.springer.com

Printed on acid-free paper

All Rights Reserved
© 2007 Springer
No part of this work may be reproduced, stored in a retrieval system, or transmitted
in any form or by any means, electronic, mechanical, photocopying, microfilming, recording
or otherwise, without written permission from the Publisher, with the exception
of any material supplied specifically for the purpose of being entered
and executed on a computer system, for exclusive use by the purchaser of the work.

I dedicate this book to my husband, my parents and my sister and brothers

Houda Laboid

Contents

Dedication	v
Preface	xiii
Foreword	xv

1
Introduction — 1

2
Wi-Fi$_{TM}$: Architecture and Functions — 5
1. WLAN Roadmap via IEEE 802.11 Family Evolution — 5
   1.1 Panorama of Standards — 6
   1.2 Features of the Different WLAN Generations — 10
   1.3 WLAN Markets — 12
2. IEEE 802.11 Architecture — 15
   2.1 Three Basic Operational Modes — 15
   2.2 Possible Configurations — 19
   2.3 Basic Services — 27
   2.4 Description of Sublayers' Structures — 27
3. Different Physical Layers — 29
   3.1 Frequency Hopping Spread Spectrum-based Physical Layer — 30
   3.2 Direct Sequence Spread Spectrum-based Physical Layer — 32
   3.3 Infrared Transmission — 35
   3.4 IEEE 802.11b Standard high Rate Direct Sequence Spread Spectrum-based Physical Layer — 36
   3.5 The OFDM Technique of the IEEE 802.11a/g Standards — 36
   3.6 The IEEE 802.11n Physical Layer — 38
   3.7 Frame Formats — 40
4. Data Link Layer — 41
5. Medium Access Control Layer — 41
   5.1 Medium Access Mechanisms — 41
   5.2 Basic Access Technique CSMA/CA — 42
   5.3 Virtual Carrier Sense CSMA/CA Mechanism with Short Messages RTS and CTS — 48
   5.4 PCF and HCF — 50
   5.5 Frames Formats — 51
6. Functions — 56
   6.1 Addressing — 56
   6.2 Association, Reassociation and Disassociation — 57

6.3	Fragmentation and Reassembling	59
6.4	Roaming	60
6.5	Synchronization	62
6.6	Energy Conserving	62
6.7	Management Frames Format	63

7. Mobility 64
8. Security 64
9. IEEE 802.11 Family and its Derivative Standards 65
   - 9.1 IEEE 802.11g 65
   - 9.2 IEEE 802.11e 66
   - 9.3 IEEE 802.11d 66
   - 9.4 IEEE 802.11F 67
   - 9.5 IEEE 802.11h 67
   - 9.6 IEEE 802.11i and IEEE 802.1X 67
   - 9.7 IEEE 802.11k 67
   - 9.8 IEEE 802.11j 68
   - 9.9 IEEE 802.11p 69
   - 9.10 IEEE 802.11u 70
   - 9.11 IEEE 802.11v 70
   - 9.12 IEEE 802.11r (Fast Roaming/Fast BSS Transition) 70
   - 9.13 IEEE 802.11s 72
   - 9.14 IEEE 802.11w 73
10. Wi-Fi and Other Technologies, Concurrency or Complementarity? 73

3
Bluetooth$_{TM}$: Architecture and Functions 75

1. Introduction 75
   - 1.1 SIG 75
   - 1.2 Bluetooth Does it Relate to Teeth by Any Sense? 76
   - 1.3 Applications 76
2. Architecture and Throughputs 77
   - 2.1 Architecture 77
   - 2.2 Throughputs and Versions 78
3. Physical Layer and Physical Channels 79
   - 3.1 Frequency Bands and RF Channels 81
4. Baseband Layer 82
   - 4.1 Physical Characteristics 82
   - 4.2 Addressing 84
   - 4.3 Bluetooth Packets 85
   - 4.4 *Error Control* 92
5. Link Manager Protocol 93
   - 5.1 LMP_Sniff_Req, LMP_Unsniff_Req 96
   - 5.2 LMP_Host_Connection_Req, LMP_Setup_Complete 96
   - 5.3 LMP_AU_RAND, LMP_SRES 98
6. Logical Link Control and Adaptation Protocol 99
   - 6.1 *L2CAP Connection Establishment Procedures* 101
   - 6.2 Some L2CAP Functions 103

7.	RFCOMM Protocol	104
8.	Service Discovery Protocol	104
9.	Profiles	105
10.	Host Control Interface	106
11.	Bluetooth Network Encapsulation Protocol	107
12.	Conclusion	108

4 IEEE 802.15.4 and ZigBee$_{TM}$ 109

1. General Architecture 109
2. Physical Layer 112
   - 2.1 2450 MHz Physical Layer 112
   - 2.2 868/915 MHz Physical Layer 113
   - 2.3 PDU Packet Format 113
3. MAC Layer 114
   - 3.1 Channel Access 114
   - 3.2 Energy Detection 115
   - 3.3 Active and Passive Scan 116
   - 3.4 Association Procedure 116
   - 3.5 Guaranteed Time Slot 116
4. Security 117
5. Frame Structures 117
   - 5.1 Beacon Frame 119
   - 5.2 Data Frame 120
6. ZigBee 120
7. Conclusion 122

5 WiMAX$_{TM}$ and IEEE 802.16 123

1. Introduction 123
   - 1.1 Backhaul Solutions 124
   - 1.2 Mobile Solutions 124
   - 1.3 Business Model 125
   - 1.4 Evolution of the IEEE 802.16 Group Activities 126
   - 1.5 Equipment Category and Frequency Bands 126
   - 1.6 Layers and Architecture 126
2. MAC Layer 127
   - 2.1 Automatic Repeat Request 128
   - 2.2 Scheduling and QoS 129
   - 2.3 Contention Resolution in IEEE 802.16 131
   - 2.4 Adaptive Antenna System 132
   - 2.5 Joining an IEEE 802.16 Network and Initialization 132
   - 2.6 PDUs 133
   - 2.7 PDU Organization 139
3. Physical Layer 139
   - 3.1 IEEE 802.16 SC (Single Carrier) 140
   - 3.2 The Single Carrier in Lower Bands 143
   - 3.3 OFDM Physical Layer 146
   - 3.4 OFDMA Physical Layer 148

6
Security in WLAN, WPAN, WSN and WMAN through Wi-Fi$_{TM}$, Bluetooth$_{TM}$, ZigBee$_{TM}$ and WiMAX$_{TM}$ — 153

1. Security in Wi-Fi Systems — 155
   - 1.1 Security in Wi-Fi Networks — 155
   - 1.2 Security Flaws — 161
   - 1.3 Taxonomy of Attacks — 171
   - 1.4 Diverse Solutions — 174
   - 1.5 WPA/WPA2: IEEE 802.1X and TKIP/AES — 193
   - 1.6 Synthesis — 194
2. Security in Bluetooth Systems — 195
   - 2.1 Security Architecture — 195
   - 2.2 Main Procedures — 200
   - 2.3 Security Flaws — 207
   - 2.4 Synthesis — 209
3. ZigBee Security — 210
   - 3.1 ZigBee Security Architecture — 211
   - 3.2 Mechanisms — 211
   - 3.3 Trust Centre Concept — 212
4. WiMAX and IEEE 802.16 Security — 213
   - 4.1 IEEE 802.16 MAC Security Sublayer — 213
   - 4.2 PKM Protocol — 214
   - 4.3 Traffic Encryption Key — 217
   - 4.4 Security Enhancement for Mobile Communications — 217
5. Conclusion — 219

7
Practising — 221

1. Mastering the TinyOS Platform — 221
   - 1.1 Brief Historical Notes and State-of-the-art in TinyOS — 222
   - 1.2 Basic Concepts of TinyOS and neSC — 224
   - 1.3 Case Study: Crossbow Motes — 227
   - 1.4 Experimentation and Measurements — 235
2. Practising WiMAX Equipment — 255
   - 2.1 General Objectives and Action Plan — 258
   - 2.2 Case Study: Redline WiMAX Equipment — 259
   - 2.3 Lab Experiments — 264

Appendix A: Structure of IEEE 802.11 Packets at Various Physical Layers — 281

1. Packet Format of Frequency Hopping Spread-spectrum Physical Layer (FHSS PHY) — 281
   - 1.1 Preamble — 281
   - 1.2 Physical Layer Convergence Protocol Header — 281
2. Packet Format of Direct-Sequence Spread Spectrum Physical Layer (DSSS PHY) — 282
   - 2.1 Preamble — 282
   - 2.2 PLCP Header — 282
3. Packet Format of IEEE 802.11b HR/DSSS PHY — 283

4. Packet Format of Infrared Physical Layer (IR PHY)	284
4.1 Preamble	284
4.2 Physical Layer Convergence Protocol header	284
5. Packet Format of OFDM PHY (Physical Layer of IEEE 802.11a)	284
5.1 Preamble	285
5.2 PLCP Header	286
Appendix B: IEEE 802.11 MAC Frames Structure	287
1. Different Types of MAC Frames	287
2. Management Frames	287
3. Beacon Frames	288
4. Association Frames	288
5. Reassociation	288
6. Disassociation Frames	289
7. Probe Request Frames	289
8. Authentication Frames	289
9. Deauthentication Frames	289
10. Traffic Indication Map Structure	290
11. Status Codes	291
Glossary	293
References	311

Preface

The advent of ubiquitous computing and the proliferation of portable computing devices have raised the importance of mobile and wireless networking. Recently, there has been a tremendous interest in broadband wireless access systems, including wireless local area networks (WLANs), broadband wireless access, and wireless personal area networks (WPANs). This domain is a subject of huge research and many standardization activities are undertaken throughout the world.

Based on the most recent developments in the field of wireless technologies related to WLAN, WPAN, wireless sensor networks (WSN) and wireless metropolitan area network (WMAN), this book gives a detailed description of the widespread or recently used standards like Wi-Fi, Bluetooth, ZigBee, and WiMAX. Our book aims at regrouping in a single volume up-to-date information related to these different technologies, which can be used separately or combined to provide specific applications. The emergence of these very promising systems is mainly due to great technological progress in the field of wireless communication protocols; they will also make it possible to offer a broad range of new applications in both civilian and military domains. The inherent characteristics of these systems imply new challenges. Our book deals with several relevant topics related to the evolution of these spontaneous, self-organized, or cellular-based networks. Through its seven chapters, we tackle critical problems such as the design of medium access control (MAC) and routing protocols, the support of the quality of service, the security mechanisms, the mobility/roaming aspects, etc. We preferred to follow an analysis-oriented approach which aims at drawing up complete states of the art on both technology and standards. We also present some practical aspects and highlight some trends as a stake for future standards. This book is intended for readers with knowledge of networking and protocols. The audience includes network engineers, designers, implementers, undergraduate/graduate/postgraduate students, and information systems managers.

The standards that we present are definitely based strongly on the knowledge of modulation and coding though we try, in this book, to pass over this domain which is widely tackled in many specialized books. Hence, we try to attract readers who are new to these technologies and make them aware of the challenges in the different layers. As standards are made by those who have the interest and patience to participate, discuss, and reach consensus, we also hope to create a new interest in participation so as to open and improve the technical levels of our future standards.

Wireless technology remains the most exciting area in telecommunications and networking, thanks to its continuous and very fast evolution. The book thus invites the reader for an exploring travel in order to understand the genesis of the most important and interesting systems, which will certainly change the landscape of future communications.

Houda Labiod
Hossam Afifi
Costantino De Santis

Foreword

It is a great honour for me to write the foreword for this book. The development of wireless communication applications and services has been incredibly boosted to the point that the ubiquitous communication concept has found many concrete expressions.

The Ethernet networks available in all offices since the 1980s allowed a comfortable computer resources usage thanks to their high bandwidth. But soon the laptop became the principal office equipment, and the Ethernet suffered from the necessity to connect physically, impeding employee mobility and the frequent need to move from offices to meeting rooms. While on the move, employees could not modify or consult data bases, email, etc. Ten years ago "Allways on" (always online) was an essential request of Internet users. Mobile communication networks such as GPRS, CDMA, and UMTS offered insufficient bandwidth for comfortable usage. The arrival of the wireless local area networks (WLANs) has made "Allways on" a reality in companies, public places, streets, hotels, supermarkets, and private residences.

Korea pioneered in the venture, and at the end of 2003, 16,000 WLANs were available for public use. Korean customers could pay a fixed price for subscription to high-speed network access, which included access to their residence by Ethernet or ADSL and connectivity through WLAN to the whole territory when on the move. In Europe hotel chains, airports, museums, metros, etc., equipped themselves to offer this service to their customers. High bandwidth access anywhere anytime is perceived today as a natural convenience associated with many service benefits (e.g. housing, travel, lecture, tourism).

The development of PDAs, game consoles, smart phones, household appliances, and cars with digital services additionally creates the need for local communication between these objects in order to allow them to cooperate. A good example is ADSL home gateways that integrate wireless access (Wi-Fi, Bluetooth, etc.) to provide easily installable triple-play services (telephony, Internet, and television) in the home. Furthermore, an accepted phone call on your mobile can pause the program on the nearest television set, use its loudspeakers, and perhaps use the television screen for a video call or a video clip received in an MMS.

Most of these objects will have access to several interfaces of communications (GPRS, Wi-Fi, Bluetooth) in order to be able to combine the services offered on every domain. The next step is the "always best connected" concept, assuming that a single object could discover the most favourable network and use it transparently for the service required at any time and in any location. This is also becoming a commercially available service in Korea and France. For instance, the telecom operator Orange has recently put on the market its "Unik" facility, which allows mobile phone users Wi-Fi access without service disruption whenever available, with associated low fares, instead of UMTS or GSM used outdoors.

Future services that will be available are body area networks (BANs), in which sensors and actuators need to communicate to send their information and to adapt their behaviour to the physical environment. Context-specific applications, such as handicapped persons' services, for instance, require new LANs that use energy sparingly, and can efficiently discover and communicate with transient neighbours such as RFID, biosensors, and biochips. New, currently designed services will need such properties. Billions of such sensors, captors, and activators will surround us and our personal assistants should be able to get the information across in time to provide us with useful enhanced services. Many of those tiny objects will have at their disposal very low energy capacity, and the choice for the optimal system design is still an open issue.

This book provides a comprehensive description of IEEE802.11, also called Wi-Fi, Bluetooth, WiMAX, or ZigBee. In-depth descriptions of the protocols are provided. The authors possess a solid practical experience of these networks that appears clearly in the explanations and numerous outlines in the text. Security, service discovery, and operating system integration are critical issues discussed in this book.

The authors have developed the important points of wireless networks describing the profits and technical elements of each approach, and explaining the integration of the WLAN in the context of the Internet.

Professors Hossam Afifi and Houda Labiod should be congratulated for their synthetic and very complete work.

Pierre Rolin
Dean of Telecom INT
Evry, France

Chapter 1

Introduction

Today wireless is becoming the leader in communication choices among users. It is not anymore a backup solution for nomadic travellers but really a new mood naturally used everywhere even when the wired communications are possible. Many technologies evolve then continuously, changing the telecommunication world. We talk about wireless local area networks (WLANs), wireless personal area networks (WPANs), wireless metropolitan area networks (WMANs), wireless wide area networks (WWANs), mobile ad hoc networks (MANETs), wireless sensor networks (WSNs) and mesh networks. Since we can find today a multitude of wireless technologies we decided to group a number of complementary technologies into one document to make it easier for a reader to understand some of the technical details of each media. Our attention has hence gone towards the most popular wireless techniques nowadays used at different scales. The first scale is the new defined "human body" scale where we envision to have many wireless devices not necessarily powered by anything, but just with ambient noise or with human heat and radiating digital information proper to ones very intimate life such as medical care.

A slightly larger scale of use cases, considered to be very important, addresses personal communications. Here, we talk about a personal "bubble" where everything belongs to a user with constraints on bandwidth and security.

In these two situations we consider in this book the available standards that fit to the usage, i.e. Bluetooth and ZigBee. Note that ZigBee is normally known under the name of Institute of Electrical and Electronics Engineers (IEEE) 802.15.4, whereas ZigBee is more targeted to design a related architecture for upper layers. In both standards a special care has been taken with power consumption. Of course with time, one can conclude that still an effort can be made over these solutions to save more energy, but every standard is related to a period where research advances are not yet mature enough to propose them as technical solutions. Even more, it may happen that a theoretical solution is not feasible with today's state-of-the-art integration and silicon technologies.

When we expand the "bubble" to larger scales with interpersonal collaborations, small office and home applications, we cannot neglect the wireless fidelity (Wi-Fi) standard of course. The larger part of this book is dedicated to that huge group of standards. Wi-Fi is going through the same growth as the Internet Protocol (IP), that makes it a great success and that takes it away from its original purpose to what people want to do effectively with it. Wi-Fi or IEEE 802.11 for technical people is an always evolving standard. We see that engineers still show an increasing interest in proposing new ideas that make better solutions. The group that meets every two months all over the world in a pseudorandom walk, is composed of a few hundreds of engineers and researchers, considered as experts in wireless and mobile communications. Contrarily to other standard bodies, that are more driven by business, there is still room in IEEE 802.11, for research studies and the pseudo-democratic procedure that is used in this area to select proposals still leaves the space for new ideas to come up. Of course, contributing to that group is still not easy as one could imagine, as one must physically follow all meetings and compete with large industrial well-known groups that control the floor.

The Wi-Fi group as we show in the book consists of a few tens of working groups each dedicated to a specific problem. The groups consider very diverse problems such as increasing the bandwidth or range, securing connections, mesh networking, vehicular communications, cooperative transmission and so on.

The last area that we consider in this book is about metropolitan radio transmission. We focus of course on IEEE 802.16 or what is commercially called worldwide interoperability for microwave access (WiMAX) as it seems to be the leading choice in this area.

IEEE 802.16 is also a very dynamic group that is competing with different bodies such as the Third Generation Partnership Project (3GPP) and even with Wi-Fi when it is deregulated in terms of transmission power.

We consider in this book two additional aspects: security and practical deployment because we believe that they are necessary to have a better view of how things are used and are put together in a real environment.

We do not consider, however, regulatory problems although it is *in se* a group in the IEEE 802 since it is strictly a per country political issue too complex to treat in a technical book. We need, however, to raise the reader's attention to the fact that regulatory wireless aspects do influence both the technique and the future.

A regulation can, for example, prevent a standard from being deployed in whole continent such as North America or Europe by disabling some frequency bands or by reducing the authorized power or channel width.

Note also that all these standards are dealt with in the IEEE 802 group that means that they are compatible with the general two layers architecture defined there. This also means that addressing, bridging and interface with upper layers are somehow the same in all these technologies.

An additional effort is done to be able to switch from one technology to the other in a seamless way. Although the group is not detailed in the book we have to present it in the introduction as a global virtual layer that enables the handover from a technology to the other without a service interruption. This effort is jointly led in IEEE 802.21 (media independent handover [MIH]) and in the Internet Engineering Task Force (IETF), where an orthogonal layer is designed to be able to extract radio information from the physical layer and to send it to an appropriate decision entity in the terminal, in the network or both to decide whether it is appropriate to change the connection from one wireless technology to the other. In the book we start with the IEEE 802.11 standard and all of its variants.

We continue with IEEE 802.16 group of standards with its different physical layers. The Bluetooth technology is described afterwards and we finish with the ZigBee standard. As explained before, two additional chapters describe the security issues in these different standards and some practical experiments that we led to provide a practical and critical view of potential usage of these technologies.

Organization. After the introduction (chapter 1), chapter 2 briefly gives a general description of WLANs. It starts by describing the context of this new type of networks within the framework of mobile networks. It then provides a rather broad outline on the particular characteristics of fourth-generation WLANs, their uses, the elements which constitute the architecture and the supported applications by quoting various types of current market segments covered by this technology. The second part of this chapter is dedicated to IEEE 802.11/802.11b (Wi-Fi) standards, which describes in an exhaustive way the mechanisms developed, with a special focus on access protocols. Standardization activities for the IEEE 802.11 family specifications are given in order to show the rapid evolution of the technical enhancements related to this technology.

Following the same clear and descriptive approach, the specification of the Bluetooth standard is detailed in chapter 3, while reviewing its major characteristics for lower and higher layers.

Chapter 4 covers IEEE 802.15.4 and ZigBee which are concerned with wireless sensor networks. ZigBee technology enables the coordination of communication among thousands of tiny sensors, which are very low cost, with lower power consumption and low data rate.

In chapter 5, we provide an overview of the IEEE 802.16 architecture and services and then look in more detail at the IEEE 802.16 specification.

Chapter 6 is devoted to security, which represents one of the major issues when deploying a wireless network. It comprises four parts. The first part includes an exhaustive state of the art of security for Wi-Fi systems, with an approach that consists in reviewing the developed mechanisms for authentification and encryption. Then we analyze thoroughly security flaws, showing an up-to-date classification of attacks. A report on standardization efforts, as well as short-term and medium-term solutions, is explicitly examined by highlighting still unsolved problems. In a similar way, the second part comprises a security study of Bluetooth systems. The third and fourth parts address, respectively, the security mechanisms related to WiMAX and ZigBee systems. A general conclusion ends this chapter by giving hot lines to set up short- and long-term security solutions.

The goal of chapter 7 is to illustrate by practical examples and experiments real WSN setups and WiMAX fresh equipment. The material is organized giving first a brief overview of TinyOS, an open source project devoted to network embedded technology and devices. Then some experiments and results are presented in a step-by-step fashion, to provide suggestions and feeling with WSNs. With reference to WMANs, the main focus is on configuration and testing of a complete wireless system composed of a base station and a couple of subscriber units using certified WiMAX equipment. A lab set-up for basic network performance measurements is showed, together with results attained by using popular tools.

The reader will find at the end of the book two appendices, which will provide him specific details concerning some important elements. Appendix A describes the structure of packets transmitted, according to the selected IEEE 802.11 physical layer. Appendix B contains a detailed description of the structure of IEEE 802.11 medium access control (MAC) frames.

Given the vast amount of information and the diversity of concepts, we preferred, in order to facilitate the reading of the book, to develop an extended glossary, which expands acronyms and provide a concise explanation of technical terms. A great number of figures are also included to provide an easy illustration of the various developed schemes as well as protocols and structures. A variety of useful documents and relevant websites are given to provide information related to the topics of this book.

This book, through its synthesis-based approach, can thus constitute an invaluable help for a broad range of readers who will benefit from an understanding of wireless communications and the associated technologies. This includes students, professionals, designers, implementers and managers.

Chapter 2

Wi-Fi$_{TM}$: Architecture and Functions

In just the past few years, WLANs have gained a tremendous place in the local area network (LAN) market. Today, WLANs based on the IEEE 802.11 standard constitute a practical and interesting solution of network connection offering mobility, flexibility, low cost of deployment and use. The purpose of this chapter is to describe in a detailed way the concepts, principles of operation and rationales behind some of the features and/or components of the standard which was originally started in 1997. We chose a clear and concise style of writing to facilitate the understanding of the inherent concepts in this technology, which sometimes require knowledge from several disciplines. Different aspects are covered throughout the chapter. Section 1 is devoted to a brief description of the main characteristics of WLANs. Then a detailed analysis of the IEEE 802.11 standard is given, including the physical transmission layer (section 3), the access to the medium (section 5) and the management functions such as addressing, association/disassociation, roaming and energy-conserving (section 6). In this chapter we focus particularly on the IEEE 802.11 and 802.11b (labelled Wi Fi) standards, nonetheless an overview of several derivative standards and working groups is shown in section 9.

1. WLAN Roadmap via IEEE 802.11 Family Evolution

Wi-Fi standard has gained some age and hence some maturity. It is still a very active group of standardization and one can see that new working or study groups are created to explore new directions. Note also that different national regulatory boards strongly influence the standard. We can see, for example, that in Europe, after the liberation of some frequencies in the 5 GHz band with good output power (1 W) makes the Wi-Fi a challenger to other long-range technologies such as WiMAX that is explained later in chapter 5.

1.1 Panorama of Standards

A WLAN is a data transmission system designed to ensure a connection that does not depend on the location of peripherals using wireless links rather than a cabled, fixed infrastructure. In companies, WLANs are generally deployed as the final link between the existing wired network and a group of customers' computers, offering a wireless access to the whole of the resources and services of the corporate network, in one or more buildings (sites).

In these last years WLANs are also getting their way into university campuses and public zones such as railway stations and airports, allowing any individual equipped with a portable computer to reach public information services or to connect itself to the Internet through the wireless infrastructure.

The widespread of WLANs depends closely on the developed standards. Indeed, the standardization ensures the reliability and the compatibility of products from different equipment suppliers. The IEEE 802 Committee, acknowledged as the world authority for LANs, defined several LAN standards during last 20 years, including the IEEE 802.3 Ethernet, the IEEE 802.5 Token Ring and the IEEE 802.3u 100BASE-T Fast-Ethernet. In 1990, the IEEE 802 Committee formed a new working group, the IEEE 802.11, specifically devoted to WLANs, with a charter to develop a physical medium specification and a MAC protocol. The Institute of Electrical and Electronics Engineers (IEEE) ratified the 802.11 specification in 1997. This standard, in its first version, featured data rates of 1 and 2 Mbps and defines the fundamental rules for signalling and wireless services. The major problem, which limited the initial industrial development of WLANs, was then the limited throughput, too low to really meet the various enterprises needs. Conscious of the need for increasing this rate, the IEEE defined the 802.11HR specification (also named IEEE 802.11 high rate or 802.11b) with speeds of 5.5 and 11 Mbps. The IEEE also worked out the specifications of the 802.11a version in the 5 GHz band.

The statutory organizations and suppliers' alliances adopted this new high data rate standard, and many market sectors appeared (Small Office Home Office [SOHO], as well as public/governmental and private/corporate related).

Apart from the standard bodies, the main players of the wireless industry met within the Wi-Fi Alliance, previously called Wireless Ethernet Compatibility Alliance (WECA). The mission of the Wi-Fi alliance is to certify the interworking and the compatibility between IEEE 802.11HR network equipments and also to promote this standard.

The Wi-Fi alliance regroups manufacturers of semiconductors for WLANs, hardware suppliers and software providers. Among them we can find companies like 3Com, Cisco-Aironet, APPLE, Breezecom, Cabletron, Compaq, Dell, Fujitsu, IBM, Intersil, Lucent Technologies, No Wires Needed, Nokia, Samsung, Symbol Technologies, Wayport and Zoom.

Up to now, the IEEE 802.11 family of standards did not cease to continuously come up with new proposals of standards. Right below we provide a list of IEEE Standards and Task Groups (TGs) that exist within the IEEE 802.11 working group. The Official 802.11 WG Project Timelines can be found at http://www.ieee802.org/11/802.11_Timelines.htm.

As it can be seen from the IEEE 802.11 home page, it is very easy to become an active member in the group where a minimum of three physical meetings are required for a person to become a voting member and hence influence the decisions taken there.

The most prevalent WLAN protocols are those related to the IEEE 802.11, IEEE 802.11b (Wi-Fi) and IEEE 802.11g standards. This family of standards deals with the physical and data link layers as defined by the OSI basic reference model (ISO/IEC 7498-1:1994). Today we see that the IEEE 802.11n is slowly taking over them.

The term used for IEEE 802.11b-certified products is Wi-Fi. Wi-Fi certification is provided by Wi-Fi alliance; notice that at present it has been extended to include IEEE 802.11g products as well. Wi-Fi alliance has also developed a certification procedure for IEEE 802.11a products called Wi-Fi5.

The list of standards and those derived from Wi-Fi are quoted below:

- IEEE 802.11: the original 1 and 2 Mbps, in the 2.4 GHz industrial, scientific and medical (ISM) band, and infrared (IR) standard (1999).

- IEEE 802.11b: enhancements to IEEE 802.11 to support 5.5 and 11 Mbps (1999).

- IEEE 802.11a: the IEEE 802.11a standard operates in the 5 GHz band and allows throughputs from 6 to 54 Mbps.

- IEEE 802.11g: allows to reach higher data rates (54 Mbps, identical to IEEE 802.11a) in the 2.4 GHz band. The orthogonal frequency-division multiplexing (OFDM) modulation is used. It provides backwards compatibility with 802.11b (2003).

- IEEE 802.11d: international (country-to-country) roaming extensions (2001), access points (APs) communicate information on available radio channels and acceptable power levels, according to countries' lawful restrictions.

- IEEE 802.11c: bridge operation procedures, included in the IEEE 802.1D standard (2003).

- IEEE 802.11e: enhancements (2005), standard for the quality of service (QoS), which defines the specifications of the QoS mechanisms to support multimedia applications. Apply to IEEE 802.11b/a/g. It introduces the hybrid coordination function (HCF). HCF uses both a contention-based channel access method, called the enhanced distributed channel access (EDCA) and a contention-free channel access method, called HCF-controlled channel access (HCCA) which have been derived from their earlier versions enhanced distributed channel function (EDCF) and HCF.

- IEEE 802.11F: deals with the standardization of protocols between APs to allow the use of a multivendor infrastructure avoiding proprietary standards. The Inter-Access Point Protocol (IAPP) offers this interworking feature.

- IEEE 802.11h: spectrum managed IEEE 802.91a (5 GHz) for European compatibility (2004). Mechanisms of frequency dynamic selection and transmit power control (TPC) are considered.

- IEEE 802.11i: enhanced security (2004). Apply to standards IEEE 802.11 b/a/g.

- IEEE 802.1X standard: provides security mechanisms for various media including wireless links by the means of strong authentication procedures with dynamic key distribution.

- IEEE 802.11j: convergence of the American (IEEE 802.11) and Japanese standards (it is the adaptation of the former to the Japanese legislation).

- IEEE 802.11k: radio resource measurement (RRM) enhancements; it defines methods and measuring criteria needed by higher layer protocols to fulfill management and maintenance functions.

- IEEE 802.11n: higher throughput improvements; it offers higher data rates (108–600 Mbps) in the 2.4 and 5 GHz bands.

- IEEE 802.11p: wireless access for the vehicular environment (WAVE).

- IEEE 802.11r: fast roaming.

- IEEE 802.11s: mesh networking.

- IEEE 802.11T: wireless performance prediction (WPP) – test methods and metrics.

- IEEE 802.11u: interworking with non-802 networks (e.g. cellular).

- IEEE 802.11v: wireless network management.
- IEEE 802.11w: protected management frames.
- IEEE 802.11y: 3650–3700 operation in the United States.

Note – there is no standard or task group named "802.11x". Rather, this term is used informally to denote any current or future IEEE 802.11 standard, in cases where further precision is not necessary. (The IEEE 802.1X standard for port-based network access control, is often mistakenly called "802.11x" when used in the context of wireless networks.)

The evolution of IEEE 802.11 standards is illustrated in Figure 2.1 which includes two types of systems: those operative in the band of 2.4 GHz and those operative in the band of 5 GHz.

FH: frequency hopping spread spectrum (FHSS) technique; DS: direct-sequence spread-spectrum technique; HR: high rate; BRAN: European project (Broadband Radio Area Network); H1: Hiperlan 1 European standard specified by the European standardization organization ETSI;

H2: Hiperlan 2 European standard specified by the European standardization organization ETSI. H2 has similar physical layer properties as IEEE 802.11a because it uses OFDM in the 5 GHz band. The MAC layer is different since it is based on a TDMA approach.

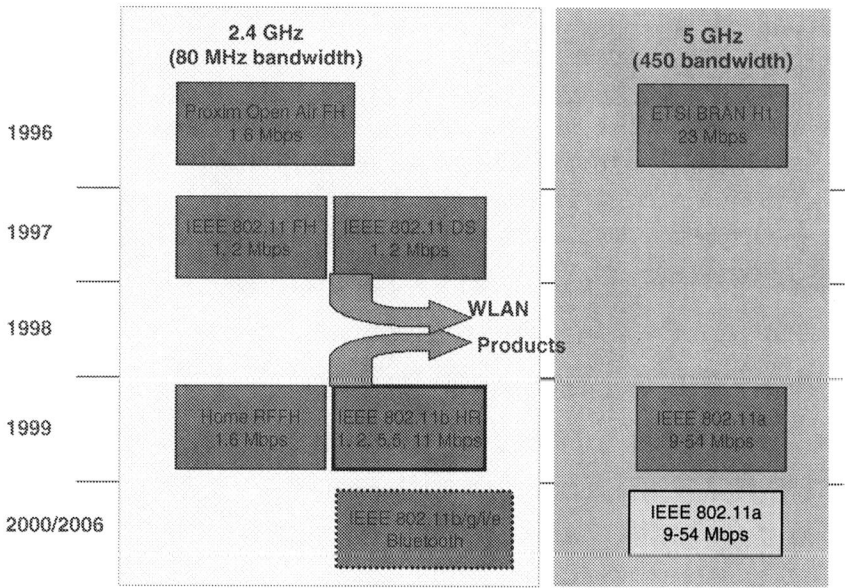

Fig. 2.1. Evolution of WLAN standards

1.2 Features of the Different WLAN Generations

A WLAN is an interesting technology because it offers a vast range of applications, thanks to its several advantageous characteristics including high capacity, short-distance coverage, full connectivity and broadcast capability. The following are the main features of WLANs.

- Licence-free operation; connection to backbone LAN
- World availability according to standards
- Theoretical throughputs definitely higher than those offered by 3G standards as universal mobile telecommunications system (UMTS)
- Low-cost design with equipment prices in progressive reduction (APs and wireless cards); very competitive costs compared to third-generation mobile systems (license + equipment)
- Roaming/handoff support
- Ease of deployment

The characteristics of these systems thus knew an evolution which can be summarized as follows:

* First generation (IEEE 802.11) since 1997 (WLAN/1G):

- Connectivity of PC terminals (between them or to a fixed LAN).
- Bridge-based APs.
- Roaming.
- Coexistence with other networks (e.g. WLAN and Ethernet LAN) which means bridging. Note that there is a small problem in the IEEE 802.11 in general with respect to bridging where it does not fulfill completely the bridging rules and is hence nonconformant to the 802 paradigms.

* Second generation (IEEE 802.11b) since 1998 (WLAN/2G):

- More effective management of WLAN
- Interworking and interoperability
- Migration starting from the first generation
- Conformity to the IEEE 802.11b standard

* Third generation (802.11a/g) since 2000 (WLAN/3G):

- High throughput (HT)
- Design of networks more open and integrated

- Conformity to the IEEE 802.11a/g standard
- Minimization of antenna sizes
- Improvement of receiver's sensitivities

* Fourth generation (IEEE 802.11n) (WLAN/4G):

  • Very high throughput (some hundreds of Mbps)
  • Long distances at high data rates (equivalent to IEEE 802.11b at 500 Mbps)
  • Use of robust technologies (e.g. multiple-input multiple-output [MIMO] and space time coding).

We compared the price of an IEEE 802.11b-based solution with that of an equivalent cabled Ethernet-based solution. The IEEE 802.11b cards used are of type Cisco-Aironet. We have three APs and 50 wireless cards (25 PCMCIA and 25 PCI cards). The total price of the equipment was almost the same. Although the initial investment is important as it is shown by the carried study, it is important to point out the immediate beneficial return of the WLAN solution (Figure 2.2). Indeed, the prices are quite similar at the beginning of the installation of the two networks, but during their lifespan, and considering also the flexibility in connectivity and in configuration changes, an IEEE 802.11b WLAN-based solution will provide more advantages. It should also be mentioned that the prices of wireless equipment is continuously decreasing nowadays.

In addition, an IEEE 802.11b solution is more advantageous than a wide area network (WAN) operator solution, because there are no specialized connections. Physical limitations are also removed and the deployment is simple and fast.

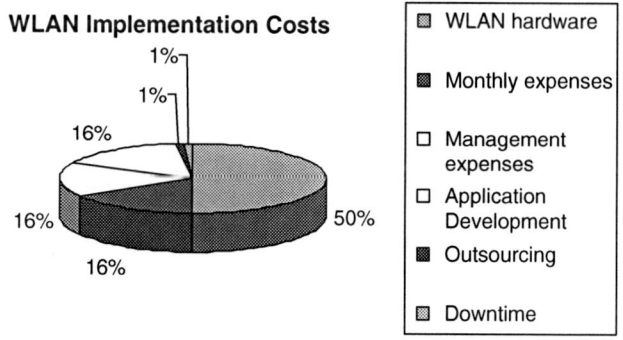

Fig. 2.2. Distribution of the costs of a WLAN solution (souce WLANA). This figure illustrates clearly that a great part of the costs relates to the used physical material (1: downtime, 2: expenses, 3: management, 4: development of applications, 5: outsourcing, 6: material)

Limitations. Independently of legal aspects, some limitations of this technology still remain such as:

- Lower data rates compared with those of high-speed fixed networks
- A limited range and influence of fixed obstacles especially metallic walls
- A shared bandwidth without a high degree of control
- Security attacks
- A quality of transmission depending on the environment (multipath propagation, path loss)
- Interworking
- Deployment
- A lack of QoS control

In order to mitigate these disadvantages, works are undertaken within several working IEEE groups. As these problems have been addressed, the popularity of WLANs has grown rapidly.

1.3 WLAN Markets

We distinguish several categories of systems where a WLAN can be used (Figure 2.3):

- Wireless personal area networks
- Wireless area networks
- Wireless metropolitan area networks

WLAN services evolve into three main categories of market segments.

Private Segment. We list three types of networks.

Enterprise networks – professional private use. WLAN solutions suitably meet a strong need for nomadism intra- and intersite for users. It remains to address two requirements: the security of access and issues related to interferences and interworking. It is a market segment which is typically filled by system integrators (deployment of WLAN networks, integration of existing systems and development of security solutions) and equipment suppliers. Many enterprise-class organizations have deployed WLAN technologies also in an attempt to increase their productivity.

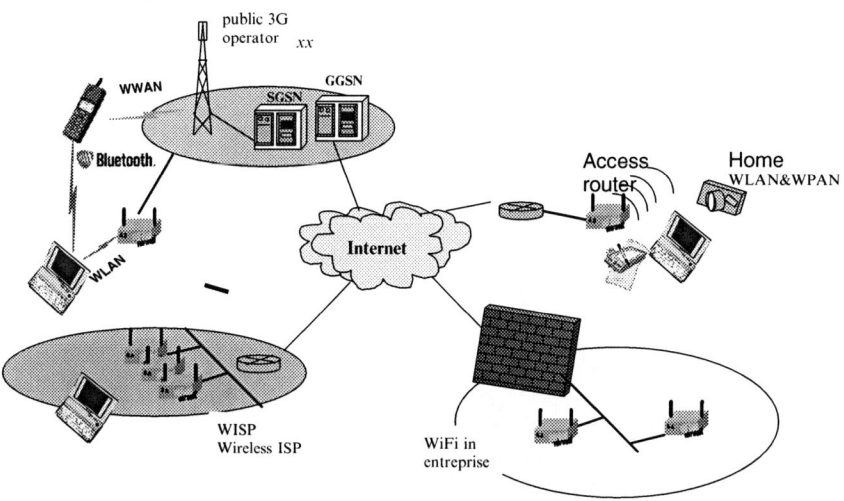

Fig. 2.3. WLAN, WPAN, WWAN: total ubiquity

Domestic wireless link – personal/home use. This kind of networks is typically deployed in a radius of a few meters around the user. The simplicity of the installation offered by a WLAN solution will probably interest, for example, a considerable part of French houses (30% of French houses are equipped with computers, 15% are multi-equipped). This solution provides free roaming inside houses, bandwidth sharing between several users, Internet access starting from an ADSL AP and it can also make radioelectric equipments communicate (house automation). Consumers of high-speed Internet access, who built their own wireless homes and small office networks, can share a fast connection among several computers.

Public private networks.

Hot Spots. Currently, it is a growing and interesting market. It concerns the high data rate access in public areas where there is an important density of users (i.e. airports, railway stations, hotels, coffee shops, libraries, subways, shopping centres, conference rooms, pirks of leisure and restaurants). A very great number of hot spots has been deployed in the United States, primarily free and cooperative (Seattle with more than 100 of relays installed to offer a free access, San Francisco, New York and Portland). In Europe, the majority of hot spots is in the Scandinavian countries and in France. The American survey firm Gartner estimated the number of hot spots in Europe to be 40,000. In the world, the annual growth is estimated, by Industrial Development Corporation (IDC), to 57%. The French operator Orange deploys 250 hot spots with approximately 1500 APs and average 200 connections per day. The directory of these hot spots is available at www.orange-wifi.com. In France, a clear tendency to the democratization of the access to the Internet appears through networks known as "libertarian" (Wi-Fi-France and Wi-Fi-Paris associations).

It is obvious that a worldwide deployment of thousands of new hot spots is well in progress and one can obtain fresher information looking for example at www.wi-fiplanet.com.

Free Networks. They are classified into two categories:

- Entirely libertarian networks which cover urban zones, although totally free, anti-terrorism laws mandate that a user still be identified for legal interception and tracking.

- Networks deployed in non-covered/non-connected rural zones.

Applications. We find several application areas for WLANs such as:

- LAN extension
- Cross-building interconnect
- Ad hoc networking
- Nomadic access
- Public hot spot access
- Data transmission within vertical markets such as:
    - LANs (offices)
    - Medical applications (real-time information transfers)
    - Retail trade, warehouse
    - Stores, shopping centres, railway stations
    - Maintenance in airports and seaports
    - Education, finance, industry
    - SOHO

Several key factors enable to foresee a rise of the WLAN market, among which we can quote:

- A progressive maturity of the standards (especially those derived from the IEEE 802.11, 802.11b Wi-Fi and 802.11a/g)

- A deployment delay of third-generation mobile systems

- A considerable investment and maturity of products (terminals, cards, adaptors, APs, etc.)

- Integration of Wi-Fi chipsets into portable laptops and tablet PCs

- Proposals for solutions of security and roaming (with certain limitations)

- An increase in the deployment of the multi-equipment in residence

- An increased mobility of users

- High throughputs of wireless systems (2G/3G/4G)

- A consequent fall in costs (Wi-Fi chips, APs, the market of PCMCIA cards disappearing)

- A frequency band usable without licence

- More widespread Wi-Fi availability

- More Wi-Fi competition

- Aggregated Wi-Fi hot spots networks (service providers wISPs, operators)

- Various markets are covered (enterprises, domestic market, telecommunication, education, health, SOHO, public sector, etc.)

We propose in section 2 a detailed technical study of the primary IEEE 802.11 standard, whose major basic mechanisms have been included in the most recent standards and more particularly Wi-Fi.

Let us note that in the continuation, the terms IEEE 802.11b and Wi-Fi will be used interchangeably.

2. IEEE 802.11 Architecture
2.1 Three Basic Operational Modes

The IEEE 802.11 standard considers two types of components: a wireless client station (in general a PC equipped with a wireless network interface card [NIC]) known as a station (STA) and an access point (AP) or sometimes called wireless relay, which functions as a bridge and a relay point between the fixed network and the wireless network. This AP is usually composed of a radio transceiver/receiver, a network card (e.g. Ethernet 802.3) and a software for layer-2 bridging in conformity with the IEEE 802.1d standard. The AP behaves like the basic station of the wireless network, aggregating the access of multiple wireless stations to the fixed network. The wireless stations include IEEE 802.11 network access cards or wireless adapters (or network interface controller). These adapters are available in many formats (PCI, PCMCIA, USB and nowadays Wi-Fi chips).

The IEEE 802.11 standard's model defines three modes: an infrastructure mode, an ad hoc mode and a mesh mode.

Infrastructure Mode. Within the infrastructure mode, the wireless network consists of at least an AP connected to the fixed network infrastructure and a set of wireless client stations. This configuration is based on a cellular architecture where the system is subdivided into cells. Each cell (called Basic Service Set[BSS], in the IEEE 802.11 is controlled by a base station (called AP).

The stations within a BSS execute the same MAC protocol and compete for access to the same shared wireless medium. We can refer to it in the following sections as a cell. Although a WLAN may be formed by a single cell (with a single AP), the maximum distance between stations is limited by many factors like RF output power and the propagation conditions of the indoor/outdoor environments. To provide for an extended coverage area, multiple BSSs are used where the APs are connected through a backbone called a distribution system (DS).

The whole interconnected WLAN including at least two different BSSs (with respect to their APs) and the DS, is seen as a single logical IEEE 802 network to the logical link control (LLC) level and is called an Extended Service Set (ESS). The majority of WLANs should be able to reach the fixed LAN services (file servers, printers and Internet access). The DS is responsible of transporting the packets between various cells within the ESS area. Data transfers occur between stations within a BSS and the DS via an AP. DS handles address mapping and interworking functions. BSSs may partially overlap. This is commonly used to cover an extended area. BSSs could be physically disjointed or co-located. To provide flexibility to the WLAN architecture, IEEE 802.11 logically separates the wireless medium from the DS medium. The DS can correspond to an Ethernet network, Token Ring, FDDI or any other communication network such as a wireless IEEE 802.11 point to point. A wide zone ESS can also provide to the various client stations an access towards a fixed network, such as Internet. Before any communication can be set within a BSS, the wireless client stations must execute an association with the AP.

Figure 2.4 shows a typical IEEE 802.11 LAN including the components described above.

The standard defines the concept of "portal or gateway", it is a device which is used to interconnect an IEEE 802.11 architecture with a traditional wired 802.x LAN. This concept is an abstract description of a part of the functionalities of a translation bridge. The portal's function is to provide a logical integration between WLAN architectures and existing wired LANs. In most hardware implementations, all the APs behave as portals. The portal logic can be implemented in a device, such as a bridge or a router or a switch, and it is a part of the WLAN and is attached to DS. In Linux, some good implementers have

Wi-Fi: Architecture and Functions

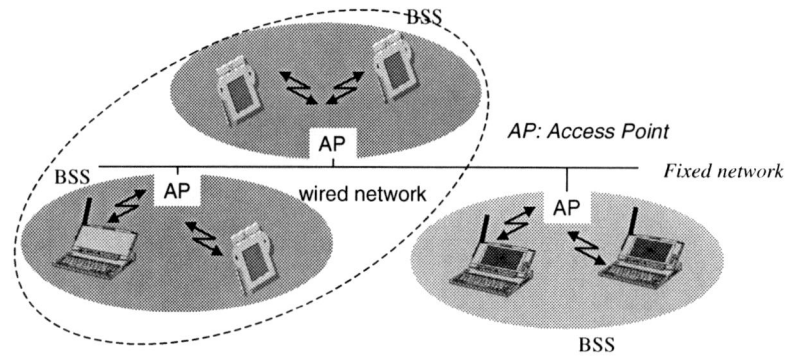

Fig. 2.4. Infrastructure mode in IEEE 802.11

provided a complete open source implementation of such a portal called *hostap*, very useful for testing any unsupported function in the current standards.

Cell identifier BSSID/ESSID and network service identifier SSID. The AP is thus the central element within a cell; since it includes the necessary functionalities to control the communications between the stations within its coverage zone with which a service zone is associated. The standard IEEE 802.11 adopts a particular addressing scheme. Each cell is identified by 6 bytes address (48 bits) called Basic Service Set Identifier (BSSID) which corresponds to the MAC address of the AP managing the cell (more details are given in the paragraph 2.6.1).

However to avoid undesirable connections, we can find an additional information called *service set identity* (SSID) that allows to identify the service network; it is a 32 bytes character string of a variable size. SSID is used in order to guarantee the authentication and the identification between an AP and a client. Any station wishing to connect itself to the extended network area must thus know as a preliminary the value of the SSID. It should be noted that this mechanism is not considered as a security protection because the identifier of the network is generally transmitted in clear through some frames (probe). So do not try to protect your network by omitting the SSID diffusion....

We should note that the infrastructure mode is the most used and it offers less performance in term of bandwidth utilization.

Nomadism and Mobility. Several types of mobility can take place (Figure 2.5):

– A local mobility within the same BSS
– A mobility intra-ESS between two different BSSs
– A mobility inter-ESS between two BSSs belonging to two different ESSs.

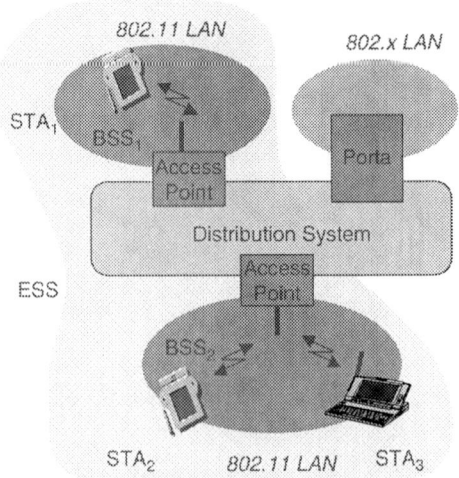

Fig. 2.5. Extended service set (ESS), distribution system (DS) and cell BSS

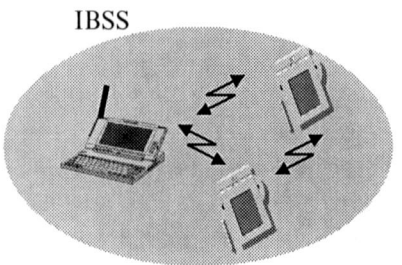

Fig. 2.6. Ad hoc mode

Ad hoc Mode. The ad hoc mode (Figure 2.6) simply represents a group of IEEE 802.11 wireless stations that communicate directly between them without having a connection with an AP or a connection to a fixed network through the DS. This configuration is sometimes referred as a peer-to-peer configuration. Each station can establish a communication with any other station in the cell which is called an independent cell Independent Basic Service Set (IBSS). These networks have been studied at the beginning of the 1970s and were named packet radio networks (PRNET). As in the infrastructure mode, an ad hoc network is also identified by an SSID identifier.

This mode allows to create quickly and simply a wireless network where there is not fixed infrastructure or where such an infrastructure is not necessary for the required services (hotel room, conference centres or airport), or finally when the access to the fixed network is prohibited or difficult.

Ad hoc wireless networks have emerged as a category of wireless networks that utilize multihop radio relays and are capable of operating in a self-organizing and self-configuring manner without the support of any fixed infrastructure. The principle behind ad hoc networking is multihop relaying, which was studied in the past under the name of PRNET in relation to defence research carried by the Defense Advanced Research Projects Agency (DARPA) in the early 1970s.

Mesh Configurations. The third type defines a hybrid configuration combining infrastructure and ad hoc modes (see section Wireless Mesh Conf.).

2.2 Possible Configurations

WLANs can meet various requirements by adopting the suitable configuration. We can distinguish several cases:

- Connection of mobile clients to the Intranet/Internet via a monocellular configuration

- Connection of mobile clients to the Intranet/Internet via a multicellular configuration

- Interconnection of distant LANs

- Interconnection of distant and isolated equipments

- Mesh networking and wireless hybrid networks

In what follows, we give a short presentation of the possible configuration cases.

Connectivity Extension to Intranet/Internet. This configuration aims at extending the connectivity to mobile users. It includes new profiles of:

- Nomadic users on a site (Figure 2.7), without a fixed office.

- Mobile users between several sites of the enterprise (or partners). Several design requirements are addressed such as security, adaptation, capacity requirements and site monitoring.

- SOHO where the network typically serves a home or a small office (no frequency reuse and probably no capacity issues).

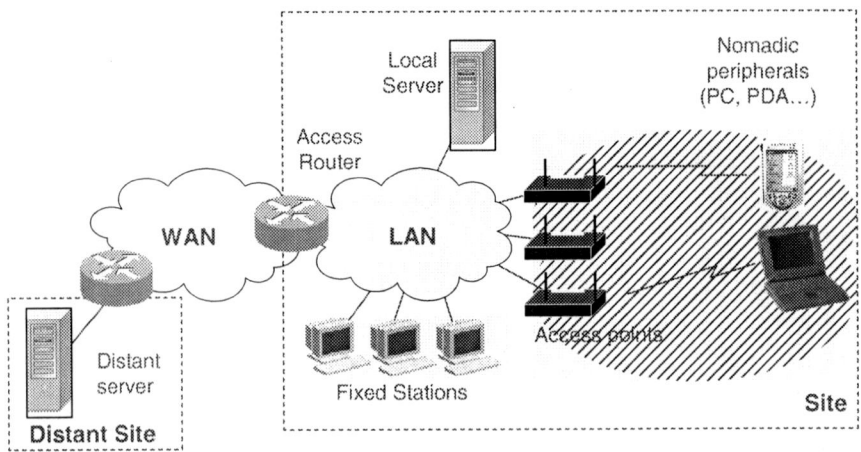

Fig. 2.7. Nomadic users – private site

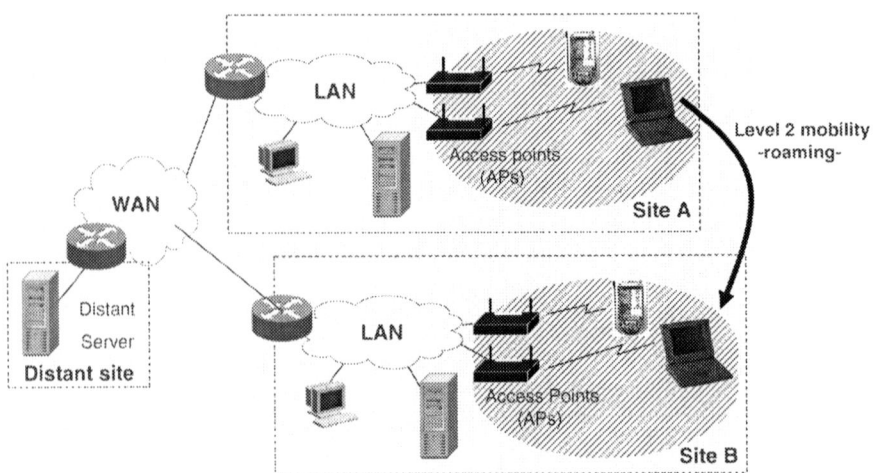

Fig. 2.8. Multicellular architecture: collaborative nomads roaming through various sites

Multicellular Configuration. This configuration uses several APs, defining distinct cells that overlap partially. In a group of contiguous cells, each AP operates on a different frequency (Figure 2.8).

Users roam in a transparent way through several sites (from a cell to another), thanks to the roaming mechanism.

Distant Network Interconnection. This configuration is a solution to extend access to structures (sites) and/or distant equipments such as:

Wi-Fi: Architecture and Functions

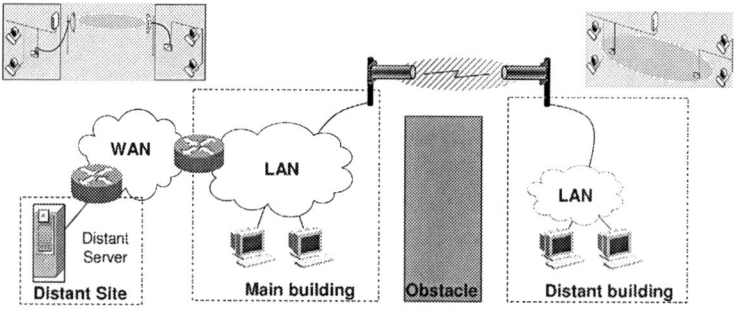

Point-to-point, point-to-multipoint links

Fig. 2.9. Interconnection of distant networks

- Buildings/equipments far away from the existing fixed network
- Buildings/equipments separated from the traditional fixed network by obstacles
- Temporary buildings/equipments

We are in the case where we use wireless bridges. We can have point-to-point or point-to-multipoint connections (Figure 2.9).

Interconnection of Isolated Equipments. This case concerns the connection of distant and isolated equipments (Figure 2.10). For example, it typically concerns a scenario that we can find in railway stations or waiting halls of airports to allow access to relevant information.

Other configurations are possible to meet potential needs of customers:

- A repeater-based configuration
- A load-balancing configuration
- A wireless switch/router configuration

Topology with Repeaters. This kind of architecture is based on the use of an AP that operates as a repeater making possible to primarily extend the covered zone (Figure 2.11).

When the extension requirements are important and the number of users is reduced, the use of this operating mode is advised. In this case, the AP just regenerates the signal and all the APs work on the same frequency. The Wi-Fi 11 Mbps are thus shared on the whole covered area between all the users.

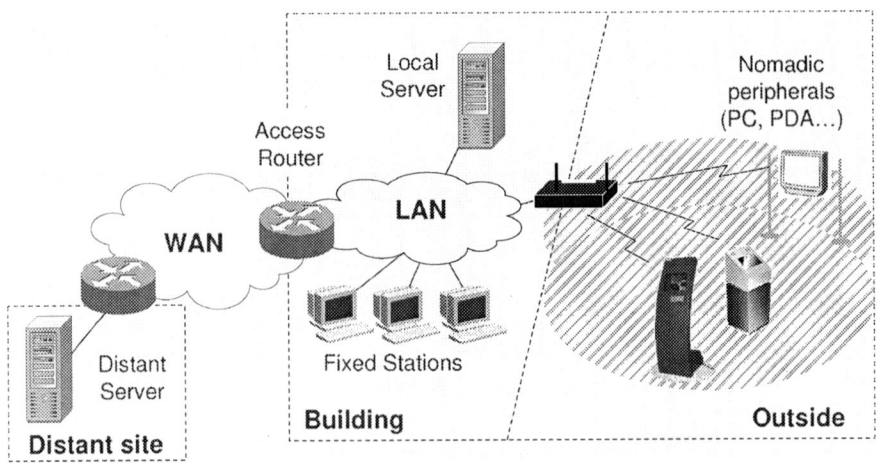

Fig. 2.10. Interconnection of fixed isolated equipment

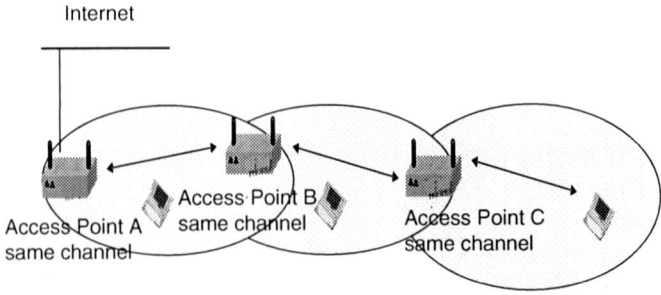

Fig. 2.11. A repeater-based configuration

Robustness and Load Sharing. By carrying out aggregations of cells, it is possible to increase the bandwidth within the same cell and to have a fault-tolerance solution if one of the available APs fails. Thus, the functionality of load balancing allows the joint use of several channels on the same zone (Figure 2.12).

To reach optimal performances, the APs being in overlapping cells must use completely disjoint frequencies in order to avoid interferences.

Wireless Switch/router-based Configuration. We can also have a configuration with a wireless switch (Figure 2.13), it is a more intelligent and centralized architecture. One or several switches are installed; they are dedicated to execute many tasks such as administration, security management and QoS management.

Wi-Fi: Architecture and Functions

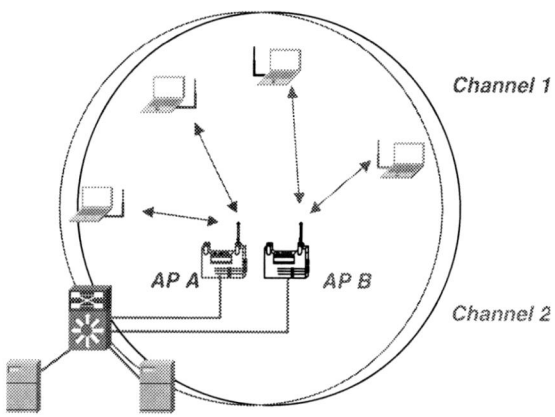

Fig. 2.12. Channel aggregation and load balancing

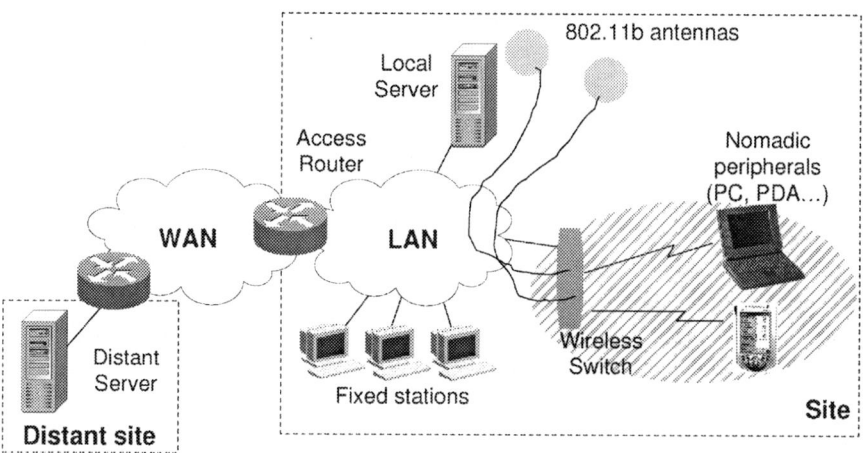

Fig. 2.13. Switch-based infrastructured configuration

Wireless Mesh Configurations. Recently, hybrid wireless networks have emerged as a promising solution, allowing mobile users to achieve service access in a seamless manner independent of their existence in communication range. In a hybrid wireless network any mobile node may have connectivity either directly or via a gateway node to an infrastructure network. This latter network may be an IP network, a 3G wide area wireless network (WAWN) or an IEEE 802.11 WLAN. Moreover, hybrid wireless networks may integrate similar or heterogeneous technologies where each mobile node moves between them in an on-demand fashion. A special case of this hybrid configuration can be a WLAN mesh network that combines ad hoc and infrastructures architectures. Some standardization efforts are paying attention to hybrid wireless networks technology. Standard organizations are actively calling for specifications for mesh networking, e.g. IEEE 802.11, IEEE 802.15, IEEE 802.16 and IEEE 802.20. The IEEE 802.11s (*targeted for June 2008*) is concerned with WLAN mesh networking.

In fact, in most WLAN deployments today, there is a clear distinction between the devices that comprise the network infrastructure and the devices that are clients that simply use the infrastructure to gain access to network resources. The most common WLAN infrastructure devices deployed today are standard IEEE 802.11 APs that provide a number of services. APs are usually directly connected to a wired network (e.g. IEEE 802.3), and simply provide wireless connectivity to client devices rather than utilizing wireless connectivity themselves. Client devices, on the other hand, are typically implemented as simple IEEE 802.11 STAs that must associate with an AP in order to gain access to the network. These simple STAs are dependent on the AP with which they are associated to communicate.

There is no reason, however, that many of the devices under consideration for use in WLANs cannot support much more flexible wireless connectivity. Dedicated infrastructure class devices such as APs should be able to establish peer-to-peer wireless links with neighbouring APs to establish a mesh backhaul infrastructure, without the need for a wired network connection to each AP. Moreover, in many cases devices traditionally categorized as clients should also be able to establish peer-to-peer wireless links with neighbouring clients and APs in a mesh network. In some cases, these mesh-enabled client devices could even provide the same services as APs to help legacy STAs gain access to the network. In this way, the mesh network extensions proposed blur the lines between infrastructure and client devices in some deployment scenarios.

Based on IEEE 802.11s definitions, a typical mesh architecture divides wireless nodes into two major classes: mesh class nodes are nodes capable of supporting mesh services, while the non-mesh class includes simple client STAs. Mesh class nodes may optionally also support AP services and may be managed or unmanaged. A WLAN mesh network topology may include mesh points (MPs) with one or more radio interfaces and may utilize one or more channels for communication between MPs.

An example WLAN Mesh is illustrated in Figure 2.14a. Any devices that support mesh services are MPs. Note that an MP may be either a dedicated infrastructure device or a user device that is able to fully participate in the formation and operation of the mesh network. A special type of MP is the mesh access point (MAP), which provides AP services in addition to mesh services. Simple STAs associate with mesh APs to gain access to the (mesh) network. Simple STAs do not participate in WLAN mesh services such as path selection and forwarding.

Wi-Fi: Architecture and Functions

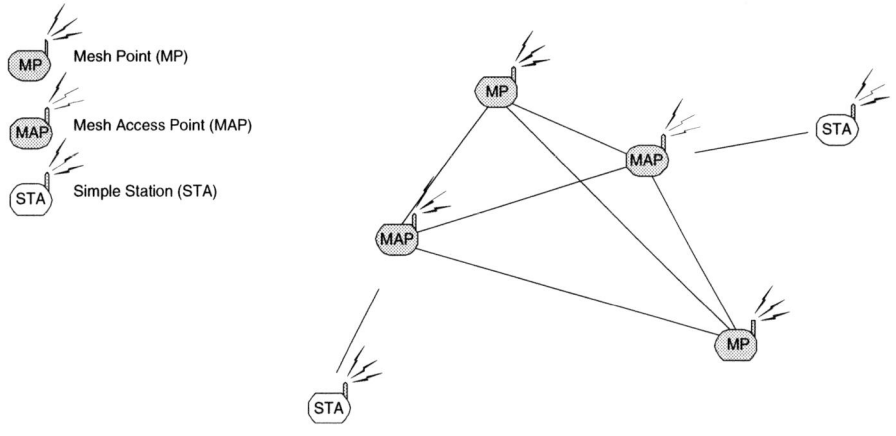

Fig. 2.14a. WLAN mesh containing MPs, MAPs and STAs

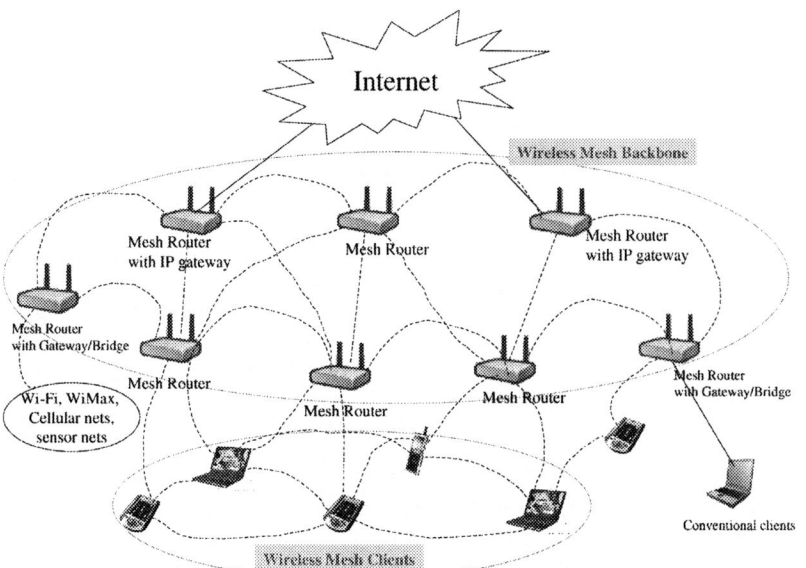

Ref: Wireless mesh networks: a survey, Computer Networks Journal, by Ian F. Akyildiz et al.

Fig. 2.14b. Infrastructure/backbone wireless mesh configuration

Thus, in principle, a single device may play the roles of both an MP and AP or the roles of both an MP and a legacy STA. These networks make use of layer-2 mesh path selection and forwarding.

Infrastructure/backbone wireless mesh networks (WMNs) are the most commonly used configurations. An example is shown in Figure 2.14b.

The main advantages of this kind of configuration are:

- Increased range/coverage and flexibility in use (compared to the basic standard)
- Reliable performance
- Seamless security
- Multimedia transport between devices
- Power efficient operation for battery operated devices
- Backwards compatibility
- Assurance of interoperability for interworking
- Possibility of increased throughput
- Density extension support
- Robustness
- Dynamic self-organization
- Self-configuration and self-healing to enable flexible integration
- Quick deployment
- Easy maintenance
- Low cost
- High scalability and reliable services
- Enhanced network capacity

Mesh networks also have significant potential disadvantages. In particular, power consumption and security are typical problems with such networking topologies. In addition, any implementation of a mesh network cannot assume that all devices will use this new protocol.

WMN is a very new topic and believed to be a promising technology playing an increasingly important role in the future generation wireless mobile networks. To achieve this aim, research efforts have been performed to propose efficient protocols and programmable configuration. However, due to the inherent characteristics, considerable works are still needed to address the problems from different layers of protocol stack and system implementation.

The MPs communicate today with two different routing protocols. The first is a reactive protocol (a protocol that fetches the route on demand and is hence reactive, i.e. slower). The second is proactive (a protocol that periodically solves all possible routes and is hence more power greedy).

2.3 Basic Services

The IEEE 802.11 architecture supports a series of basic services that are:

– Association/disassociation/reassociation

– Delivery of the MAC/MSDU frames

– Authentication/deauthentication

– Diffusion and broadcast

– Beacon and probing

– Privacy/confidentiality

– Higher-layer timer synchronization/QoS traffic scheduling

– Radio measurements

2.4 Description of Sublayers' Structures

Similarly to other IEEE 802 standards, IEEE 802.11 focuses on the two lowest sublayers of the ISO model (interconnection of the open systems), the physical layer (PHY) and a data link layer containing the MAC sublayer and the LLC) as shown in Figure 2.15.

The physical layer, which is in charge of the transmission of the MAC frames on the wireless medium, uses several modulation techniques and binary coding. The reader interested by the details of the techniques for modulation, RF design and signal processing can consult the references quoted at the end of this chapter [BRO 97, LEE 95, PRO 95, WAL 02].

The MAC layer is described thereafter by presenting its various functions (association, fragmentation, access control, security, etc.) as well as the structures of MAC 802.11 frames. The IEEE 802 standard defines the same data link high sublayer for all the LANs; it acts as a LLC 802.2 sublayer. IEEE 802.11 standard thus describes only the physical and MAC layers. We can thus lead to a total stacking of the ISO layers like illustrated in Figure 2.15a.

Notice 2.1 – Architecture, functions and basic services of Wi-Fi (IEEE 802.11b) are defined by the original IEEE 802.11 standard. IEEE 802.11b specification affects only the physical layer offering higher data rates and a more robust connectivity.

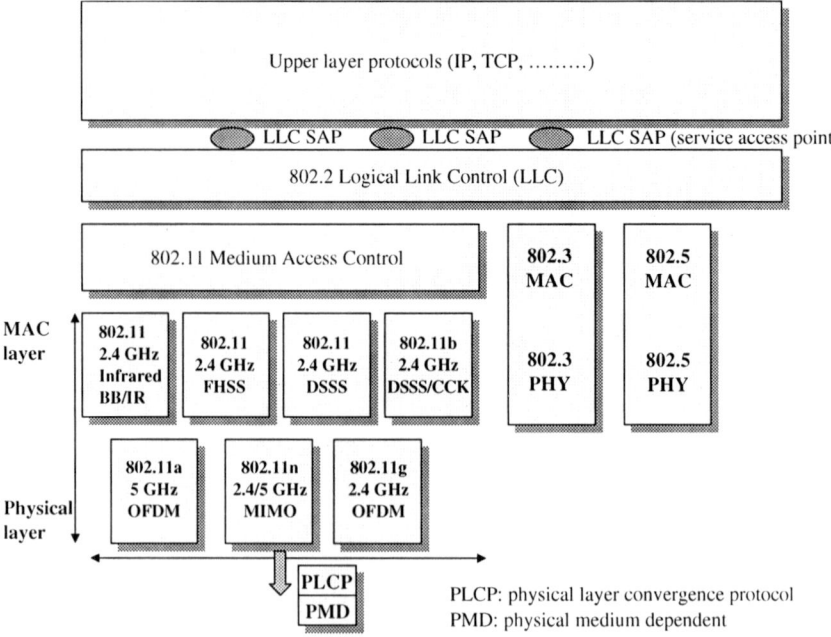

Fig. 2.15a. IEEE 802.11 and 802 family: OSI reference model

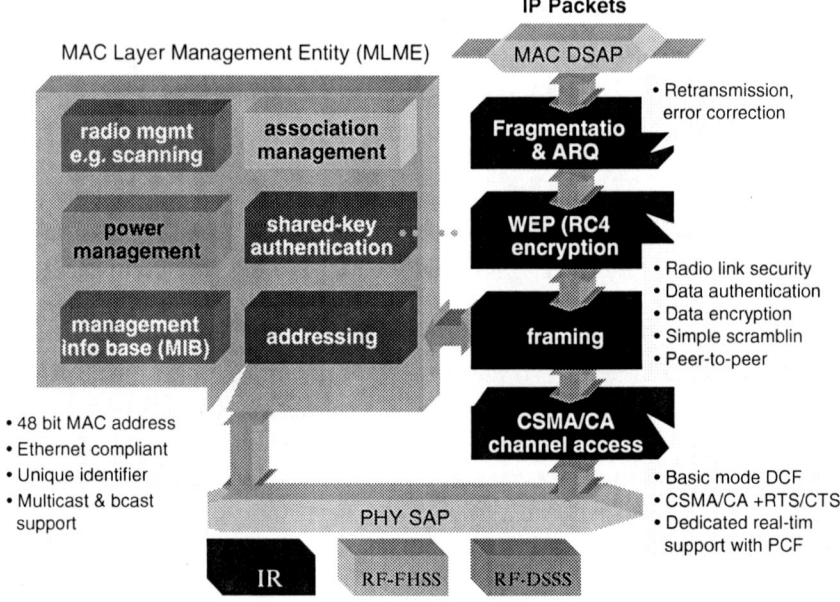

Fig. 2.15b. Global view of IEEE 802.11 MAC layer

3. Different Physical Layers

Several types of radio interfaces can be supported by the IEEE 802.11 reference architecture. Three different physical layers are defined in the basic standard:

- A radio physical layer using an FHSS technique operating in the 2.4 GHz ISM band (see paragraph 3.1).

- A radio physical layer using a DSSS technique [LEE 95, PRO 95] operating in the 2.4 GHz ISM band (paragraph 3.2).

- A physical layer for the IR transmission [BAR 94] (paragraph 3.3). Of course, there has never been any real commercial product for such a physical choice.

Only the transmission by IR operates in baseband at wavelength between 850 and 950 nm. The transmissions based on spread spectrum techniques allow increasing the performances by minimizing the harmful effects of the multipath propagation, interferences and noise. In standards 2G/3G WLAN, other modulations have been defined:

- Complementary codes keying (CCK) modulation in the case of the IEEE 802.11b system; the associated physical layer is called HR/DSSS (high rate DSSS).

- Orthogonal frequency multiplexing division (OFDM) modulation (in the case of IEEE 802.11a and 802.11g systems [NEE 00].

- MIMO techniques in IEEE 802.11n.

For IEEE 802.11, Wi-Fi (IEEE 802.11b) and IEEE 802.11g, the frequency band used is a band without licence known as Industrial Scientific Medical (ISM) around 2.4 GHz. The bandwidth available is 83 MHz. For IEEE 802.11a-based systems, a greater bandwidth is available; it is approximately equal to 455 MHz in the 5 GHz band. This band is unlicenced but provides an upper bound on power transmission usually limits to 100 m Watt.

The physical layer is structured in two sublayers:

- The convergence sublayer

- The sublayer dependent on the transmission medium itself: physical medium dependent (PMD).

The high part is called the convergence sublayer or Physical Layer Convergence Sublayer Procedure (PLCP). The purpose of this layer is to adapt to the lower sublayer which is dependent on the medium (OFDM, IR, DSSS or

FHSS). This layer has as a role to insert headers required for synchronization or identification of the used modulation on the medium. It also allows choosing the best antenna to capture the signal (in the case of an AP using the antenna diversity technique and MIMO). The frames sent by the convergence layer are called PLCP Protocol Data Links (PPDU). The format of the PPDU depends on the used lower sublayer. The procedure of PLCP transmission is invoked by the carrier sense/clear channel assessment (CS/CCA) procedure which is executed:

- To detect the beginning of a signal that can be captured (CS phase)

- To determine if the channel is free before transmitting a packet (CCA phase)

Minimal duration of the CS/CCA procedure is equal to the duration of a slot whose value depends on the characteristics of the selected physical layer (see paragraph 5.2). IEEE 802.11 does not specify the procedures to detect that the support is occupied, the equipment suppliers have thus the choice to implement methods more or less innovating while respecting the lawful constraints on the use of the frequencies (power limitations).

The lower sublayer role has primarily to code and transmit the bits sent by the convergence layer on the medium. This layer is called PMD layer.

In addition to the information transport plan, a control plan exists. All the control (management) functions related to the physical layer are implemented in a PHY management layer. Management information is stored in a database called Management Information Base (MIB), usually based on the SNMP format. A specific group in the IEEE 802.11 group (called TGV) works on the management plane.

3.1 Frequency Hopping Spread Spectrum-based Physical Layer

It offers 1 or 2 Mbps throughputs. The frequency band used is located between 2.4 and 2.483 GHz (see Figure 2.16). 79 channels of 1 MHz (35 channels in France because of exclusive use of certain channels by the defence) are available to the United States. The signal is transmitted alternatively on the various available channels. The transmitter and the receiver are synchronized on a succession of frequency hops during the communication. The sequence of hops (multiple channels) defines a pattern. The hop sequences are defined based on pseudo-random sequences for civil equipments and on a secret key for the military equipments. The change of frequency on this sequence must be made approximately every 400 ms. This dwell duration on a given frequency is called a "hop" and is defined by the regulation. A synchronization of the frequency changes is thus necessary.

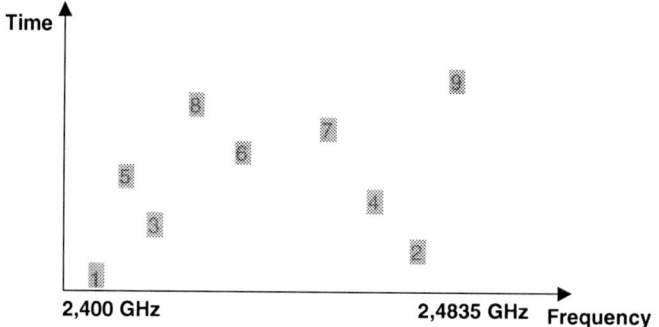

Fig. 2.16. FHSS technique

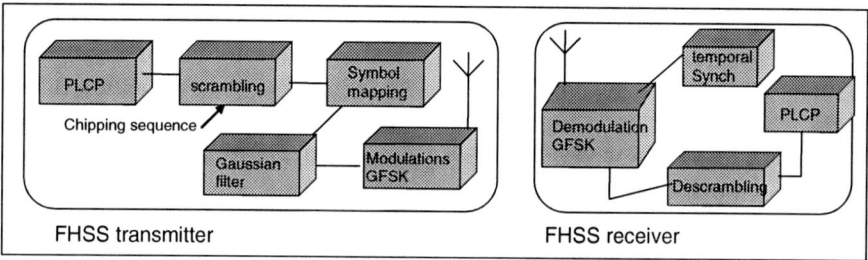

Fig. 2.17. FHSS transmitter/receiver structure

The advantages of this technique are numerous. Indeed, it prevents a total loss of the signal (only the coded signal on the scrambled carrier frequency is lost and is retransmitted generally besides on the next hop). It provides a good immunity against interferences, fading and background noise. This method thus constitutes an effective solution in a multipath environment.

The transmission by FHSS offers two throughputs. The modulations used are:

– A frequency modulation Gaussian frequency shift keying (GFSK) with two levels for 1 Mbps

– A frequency modulation GFSK with four levels for 2 Mbps

The preambles and frame headers in both cases are transmitted with a modulation of 1 Mbps.

Figure 2.17 shows the structure of an FHSS transmitter/receiver.

The existing equipment on the market supporting a physical layer with 2 Mbps can have the capacity to reduce their throughput to 1 Mbps if the quality of the signal becomes bad. This may be seen as good or bad depending on the usage. It is good because far station still can emit and receive data, while it is also considered bad because a slower station monopolizes the channel more time than a fast station slowing down the overall cell performance.

3.2 Direct Sequence Spread Spectrum-based Physical Layer

This technique uses a single frequency band contrary to FHSS technique. The used frequency band is 2.4–2.4835 GHz. 14 channels of 22 MHz separated by 5 MHz are defined. Adjacent channels partially overlap as it is shown in Figure 2.18.

In the original version of the standard, the DSSS transmission offers two rates. The modulations used are:

- A phase modulation differential binary phase shift keying (DBPSK) for 1 Mbps

- A modulation in squaring of phase DQPSK for 2 Mbps

These two modulations are based on the principle of the differential phase modulation. Thus, a new symbol to be transmitted results in a rotation of the phase of the signal. This technique thus requires to know the reference of phase only for the first emitted symbol. Then, a transmission of a symbol resulting in a rotation of the phase of the signal compared to the phase of the preceding symbol, it is not necessary any more to know the phase of reference.

In the binary frequency modulation DBPSK, only one information bit is emitted by a symbol. Thus, the appearance of one "0" does not cause rotation of phase whereas the appearance of one "1" causes an inversion of phase. In the frequency modulation in squaring of phase DQPSK, one transmits two bits of information by symbol. The rate of transmission is thus multiplied by two. The rotation of phase is summarized in Table 2.1.

More important throughputs are offered in the standard IEEE 802.11b: 5.5 and 11 Mbps. The preamble and the header of a frame at the physical layer are transmitted at 1 Mbps.

In order to avoid the problems of interferences and/or of noise, we can use a technique of chipping which can spread the spectrum. Each bit of information is

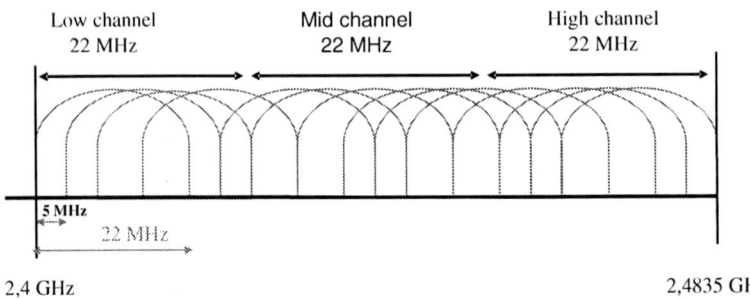

Fig. 2.18. 2.4 GHz channels

Table 2.1. Different phase rotations

Symbol	Rotation
00	0°
01	90°
11	180°
10	270°

coded in a succession of bits called chips. This sequence in general consists of 11 bits agreed upon between the transmitter and the receiver. A Barker code is used with a sequence of chipping "$+1, -1, +1, +1, -1, +1, +1, +1, -1, -1, -1$". This redundant coding permits to recover a complete bit even when a great part of the signal is lost. The transmission is carried out only on one fixed frequency contrarily to the FHSS technique.

It should be noted that this method of chipping introduces a redundancy in addition to a spread spectrum technique which makes the signal to be transmitted more robust to the errors. Indeed, if the noise affects only one part of the band, it will be possible to restore the signal and to recover the bits of information. The sequence of chips has 6 chips set to 1 and 5 to -1. Thus, this distribution makes it possible the receiving station to find its synchronization thanks to a measurement of the autocorrelation of the signal received with the sequence of chips of the receiver. It is necessary to notice that each time that a bit is emitted, the receiver multiplies its sequence of 11 chips with the sequence of the received signal: if there is a shift between the sequence of the receiver and that of the emitted signal, it is difficult to decode the signal. The measurement of the autocorrelation between these two sequences allows to determine if the receiver is in shift compared to the sequences emitted by the source.

A mechanism of throughput variation according to the quality of the wireless links is applied.

Figure 2.19 shows the structure of a DSSS transmitter/receiver which is more complex than that of a FHSS transmitter/receiver.

A French characteristic concerning the regulation for the frequencies was that until the end of 2001 only four frequencies were used because of the existence of defencee radars in the band 2.4 GHz. Gradually, a total use of the band becomes effective on the whole of the territory. Moreover it is now free to use 5 GHz transmitters with high-power throughput, a solution that we said can be competitive with WiMAX solutions.

Given that the available channels in the 2.4 GHz frequency band overlap partially, interferences on adjacent channels can take place. However, it is possible to choose three perfectly disjoint channels. For example, in Europe, we can find

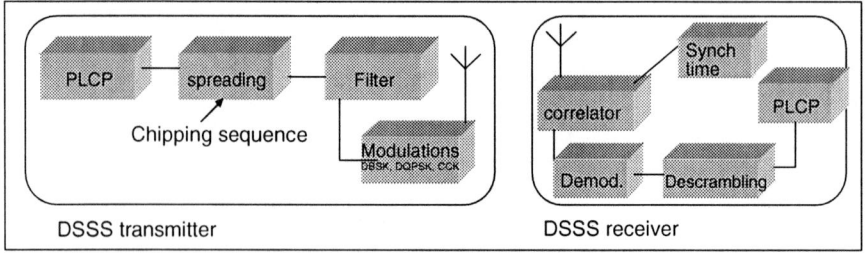

Fig. 2.19. DSSS transmitter/receiver

the channels (1[2412 MHz], 7[2442 MHz], 13[2472 MHz]) and in the United States (1[2412 MHz], 6[2437 MHz], 11[2462 MHz]). Consequently, three APs defined to cover three cells make possible to aggregate throughput and to offer a total throughput of 3 times the maximum bandwidth of the cell.

FHSS or DSSS?. The decision should be based on the customer applications' requirements. If the needs are directed towards a broad cover or High throughputs, then the selected technique will be DSSS. In addition, in the case of a highly multipath environment, it is preferable to use the FHSS technique.

The majority of the currently existing products on the market mainly implement the IEEE 802.11b DSSS standard that offers four throughputs. The selected throughput is defined dynamically according to the radio conditions. Figure 2.20 illustrates this mechanism. We notice that more the equipment is close to the AP, more the throughput used is high for both FHSS and DSSS. The first zone around the AP makes it possible to reach throughputs of about 11 Mbps whereas when one move away, the throughput falls to 5.5, then 2 and 1 Mbps. Located at the same distance of the AP, the throughput offered by DSSS is multiplied approximately by a factor of 2 compared to that offered by FHSS.

The extent of the covering zone in DSSS is generally related to the throughputs used with a distinction according to the type of the environment (indoor or outdoor) (Tables 2.2 and 2.3).

The goal of this section is to give to the reader an idea on the orders of magnitude of the ranges according to the throughputs. It should be noted that these measurements were taken in a particular environment and are thus related to the characteristics of materials of the buildings and the environments in general.

In some configurations, a total degradation of the cell in term of throughput may occur if a station is distant from the AP and chooses to transmit to a low throughput.

Wi-Fi: Architecture and Functions

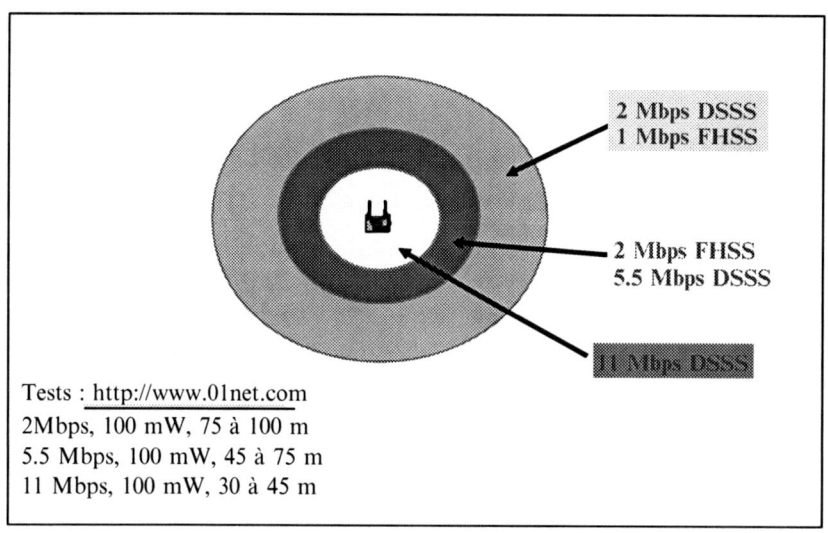

Fig. 2.20. Throughput variation

Table 2.2. Indoor environment, DSSS technique

Throughput (Mbps)	Range (meters)
11	50
5.5	75
2	100
1	150

Table 2.3. Outdoor environment, DSSS technique

Throughput (Mbps)	Range (meters)
11	200
5.5	300
2	400
1	500

3.3 Infrared Transmission

This transmission is not directive and applied to indoor environment only. Crossing of the obstacles (e.g. walls, ceilings and partitions) is not possible. The IR interfaces should not be exposed to the light of the sun, they are sensitive to snow and fog. The transmission is done with a wavelength located between 850 and 950 nm with a power emitted of up to 2 W. The modulation used is position modulation pulsates (PPM).

3.4 IEEE 802.11b Standard high Rate Direct Sequence Spread Spectrum-based Physical Layer

In addition to the 1 and 2 Mbps transmissions supported by the basic standard, transmissions with 5.5 and 11 Mbps are possible in IEEE 802.11b thanks to the use of the complementary code keying (CCK) modulation. Information is sent by using special and sophisticated complementary codes. The chipping rate is of 11 Mchip/s with an 8-chip symbol duration. The 5.5 Mbps throughput is obtained with a transmission of 4 bits per symbol instead of 8 to reach a throughput of 11 Mbps. Also, the algorithmic of CCK makes it possible the receivers to easily distinguish the various codes even in the presence of interferences and fading due to multipath. The process of generation of CCK symbols resembles in general to that used in DSSS with the difference that the code words are partially derived starting from the data; on the other hand a static code Barker is not used and the spreading of the signal is carried out using CCK symbols.

The supported MAC layer is always the same. Although the modulation used is different, the available channels are identical to those authorized for DSSS PHY layer with lower throughput. One nice experiment that one can do is to switch on two different APs with preconfigured different physical layers (usually we can set up a 11b with a 11a). The two APs are interconnected via Ethernet. When we switch off one of the APs the station handovers to the second one but all communications remain and the data flow continue to arrive through the new interface. It is called a seamless handover.

3.5 The OFDM Technique of the IEEE 802.11a/g Standards

IEEE 802.11a and 802.11g systems use the OFDM modulation in the 5 GHz/ 2.46 Hz ISM band. The total bandwidth available is approximately 455 MHz of which 200 are allocated to indoor use and 255 to the outdoor use. This modulation allows to reach throughputs between 6 and 54 Mbps and provides good performances in presence of multipath. The MAC layer is identical to that defined for the other physical layers (FHSS, DSSS, HR/DSSS, IR). The physical layer supporting OFDM modulation is complex because it applies a varied whole of techniques of digital transmission such as the phase modulation, OFDM multicarrier transmission, convolutional coding and interleaving. Let us recall that OFDM was proposed for the first time by the ETSI within the European system Hiperlan 1. OFDM is based on a frequential division where

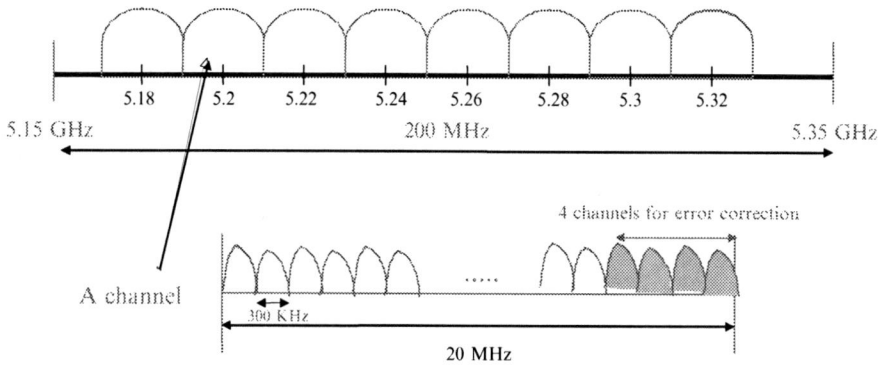

Fig. 2.21. OFDM channels in the 5 GHz lowest band

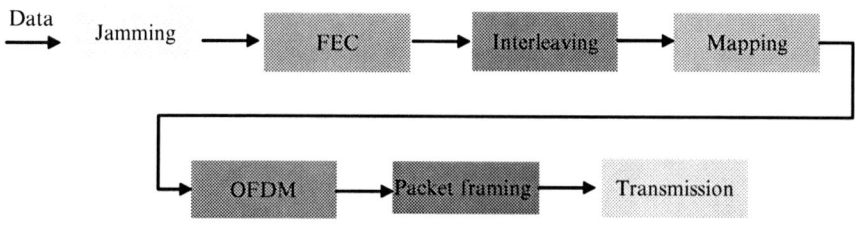

Fig. 2.22. OFDM transmission principle

the frequency band is divided into carriers which can be used simultaneously by multiplexing the data on subcarriers. A channel consists of 52 carriers of 300 KHz width. 48 carriers are dedicated to information transport and 4 for the error correction pilot carriers. OFDM supports a series of modulations and codes making it possible to offer the whole set of throughputs.

Eight channels of 20 MHz are defined in the low band (from 5.15 to 5.35 GHz). We can have a colocalization of eight networks within the same space and a maximum throughput of 432 Mbps (see Figure 2.21).

The total transmission chain is summarized by the diagram of Figure 2.22.

A first step consists in choosing the throughput according to the wireless link state; we execute in fact an automatic mechanism of link adaptation. Once the data frames are built, we proceed to a rearrangement of the data elements (jamming or scrambling). We apply a coding without backward channel forward error correction (FEC) by adding redundancy. In order to reinforce the protection of information, we carry out an interleaving of the coded bits which are then mapped in constellations of 1, 2, 4 or 6 points. Each byte represents a complex number according to the type of the used modulation. Forty-eight complex numbers are then assembled to form an OFDM symbol, a carrier is assigned to each one of these symbols. The OFDM modulation application permits to obtain a signal in baseband. The duration of an OFDM symbol is 4 μs. The packet of the physical layer is thus built by adding the necessary fields (preamble) for its transmission on the wireless medium. Note that in this part and in other chapters, information related to modulation and coding, are not really deeply explained because it is more in the field of signal processing and propagation that can be found in specialized digital com. books.

3.6 The IEEE 802.11n Physical Layer

In the N option, the real data throughput is estimated to reach a theoretical 540 Mbps (which may require an even higher raw data rate at the physical layer), and should be up to 100 times faster than IEEE 802.11b, and well over ten times faster than IEEE 802.11a or IEEE 802.11g. IEEE 802.11n will probably offer a better operating distance than current networks. IEEE 802.11n builds upon previous IEEE 802.11 standards by adding MIMO. MIMO uses multiple transmitter and receiver antennae to allow for increased data throughput through spatial multiplexing and increased range by exploiting the spatial diversity and powerful coding schemes.

An AP declares its capabilities in the HT capability information in each beacon.

The N system is strongly based on the IEEE 802.11e QoS specification to improve bandwidth performance. The system supports basebands width of 20 or 40 MHz and can be in one of the four situations:

- **Low mode** – the operation corresponds to classical a/b or g
- **HT-mode** – the device can operate in either 20 or 40 MHz bandwidth and with 1–4 spatial streams (MIMO).
- **Duplicate non-HT mode** – in this mode the device operates in a 40 MHz channel composed of two adjacent 20 MHz channels.
- **40 MHz upper mode** – in this mode the device transmits frames in the upper 20 MHz channel of a 40 MHz channel.
- **40 MHz lower mode** – We transmit a frame in the lower 20 MHz channel of a 40 MHz channel.

Table 2.4 summarizes the maximum bitrates that can be obtained with a MIMO configuration of four antennae and with a 40 MHz bandwidth.

A typical transmitter block diagram is shown in Figure 2.23.

The blocks are briefly described hereafter:

- **Scrambler** – scrambles data to prevent long sequences of zeros or ones.

Table 2.4. IEEE 802.11n bit rates

Modulation	Mbps
BPSK	60.0
QPSK	120.0
QPSK	180.0
16-QAM	240.0
16-QAM	360.0
64-QAM	480.0
64-QAM	540.0
64-QAM	600.0

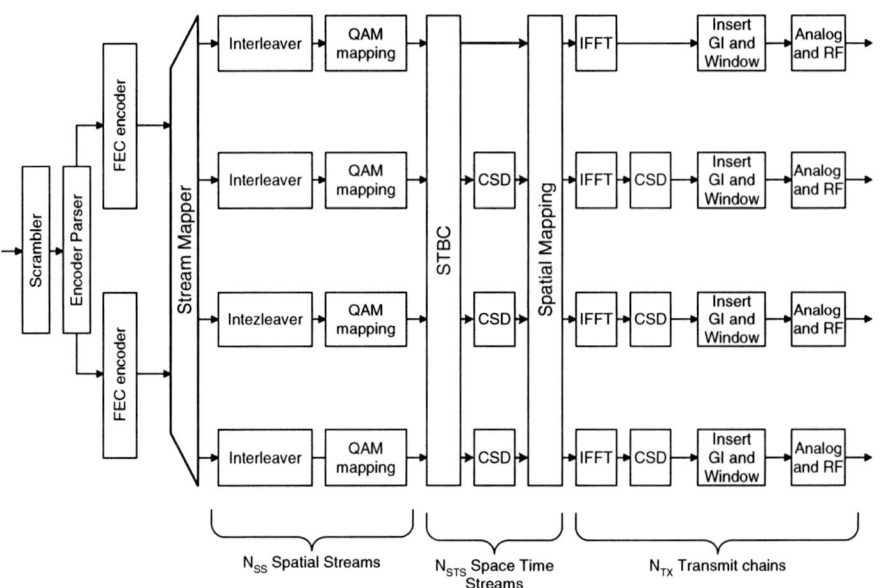

Fig. 2.23. IEEE 802.11n transmitter

- **Encoder parser** – de-multiplexes the scrambled bits among FEC encoders, in a round-robin manner.

- **FEC encoders** – encodes data to enable error correction – FEC encoders may include binary convolutional encoders followed by puncturing device, or LDPC encoder.

- **Stream parser** – divides the outputs of the encoders into blocks that will be sent to interleavers and mapping devices. The sequences of the bits sent to the interleavers are called spatial streams.

- **Interleaver** – interleaves the bits of each spatial stream (changes order of bits) to prevent long sequences of adjacent noisy bits from entering the FEC decoder.

- **QAM mapping** – maps the sequence of bits in each spatial stream to constellation points (complex numbers).

- **Space time block coding** – constellation points from one spatial stream are spread into two space time streams using a space time block code.

- **Spatial mapping** – maps space time streams to different transmit chains. This may include one of the following:

    - **Direct mapping** – constellation points from each space time stream are sent to a different transmit chain.

    - **Spatial expansion** – vectors of constellation points from all the space time streams are multiplied by a matrix to produce the input to the transmit chains.

    - **Beamforming** – similar to spatial expansion: each vector of constellation points from all the space time streams is multiplied by a matrix of steering vectors to produce the input to the transmit chains.

- **Inverse fast Fourier transform** – converts a block of constellation points to a time-domain block.

- **Cyclic shift insertion** – inserts the cyclic shift into the time-domain block. In the case that spatial expansion is applied that increases the number of transmit chains, the cyclic shift may be applied in the frequency domain as part of spatial expansion.

- **Guard interval (GI) insertion** – inserts the guard interval.

The reader who wishes to look further into the quoted digital transmission techniques finds at the end of this chapter, a very broad whole of references on the subject [BRO 97, COLLAR 99c, COLLAR 00, EDO 97, NEE 00, PRO 95].

3.7 Frame Formats

The reader will find in Appendix A the detailed structure of the various frames associated with the various physical layers.

4. Data Link Layer

It is made up of two sublayers. LLC uses the same properties that an LLC 802.2 layer and allows to connect a WLAN to any other LAN of the IEEE family. The MAC layer is specific to IEEE 802.11 and defines new mechanisms of access to the medium. It is independent of the characteristics of the physical layer and the throughputs and supports two topologies infrastructure and ad hoc. In future we will also find the mesh facility in the same layer. It also supports the following services:

- Asynchronous data transmission service (mandatory):
  - Does not offer guarantee (best-effort service)
  - Broadcast and multicast traffics support
  - Implemented using DCF mechanism (see paragraph 5.2)
- Time constraints traffic service (optional)
  - Real-time service
  - Implemented using the PCF mechanism (see paragraph 5.4)

In addition to data transmission, other basic services are provided such as:

- Association/disassociation
- Privacy (wired equivalent privacy [WEP] mechanism, etc.)
- Authentication and access control
- Fragmentation/reassembly and ordering
- Energy saving

A global view of the MAC layer is given in Figure 2.15b. The functionalities are detailed in section 6.

5. Medium Access Control Layer

5.1 Medium Access Mechanisms

The MAC layer defines two different access methods, the distributed co-ordination function (DCF) which is used for a distributed and random access like that of IEEE 802.3 with several other algorithms specific to the WLAN and another method which is the point coordination function (PCF) for a controlled access. The DCF method is defined to deal with the transport of the asynchronous best-effort traffic. All the users who want to transmit have an equal chance to access to the medium. The PCF is

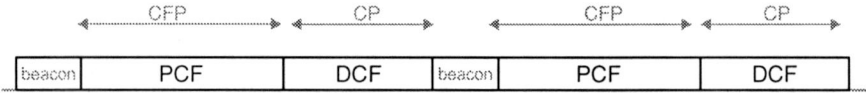

Fig. 2.24. Structure of superframe: CFP and CP periods

now replaced by the hybrid coordination function (HCF) and is based on a polling approach plus a differentiated transmission by assigning different waiting periods accord to nodes priorities. It requires the existence of an entity for control. HCF is conceived for the data transmission with time constraints such as the real-time traffic (voice, video). DCF is thus the basic access method with contention whereas HCF is an optional method without contention. Contention free period (CFP) is the HCF period and contention period (CP) is the duration of contention. Information relative to the two access modes with and without contention is disseminated in the BSS through control frames called beacon frames or beacons (equivalent to the BCCH channel in GSM cellular networks). A superframe is defined by a CFP part followed by the DCF mode part (Figure 2.24). The beginning of a superframe is delimited by a beacon frame.

In what follows, we will detail the two types of access mechanisms.

5.2   Basic Access Technique CSMA/CA

The basic access mechanism, called DCF is typically the carrier sense multiple access with collision avoidance (CSMA/CA) mechanism. CSMA protocols are well known and Ethernet is the most famous one, which is a protocol based on the CSMA/CD access mechanism (CD for collision detection). Contrarily to CSMA/CD mechanism which is based on the collision detection, the CSMA/CA allows an access to a shared medium by avoiding collisions.

Collision detection mechanisms are well suited to a fixed LAN, but they cannot be used in a wireless environment, for two main reasons:

- Implementing a collision detection mechanism would require the implementation of a full duplex radio capable of transmitting and receiving simultaneously; and we have not such terminals.

- In a wireless environment, we cannot be sure that all the stations hear each other (which is the basic assumption of the collision detection principle). Moreover, the fact that a station wanting to transmit checks if the medium is free, does not necessarily mean that the medium is free around the receiver area (multipath channels, fading).

- The carrier sense procedure takes a relatively long time in comparison with the fixed networks (typically 30–50 μs in fixed networks).

Wi-Fi: Architecture and Functions

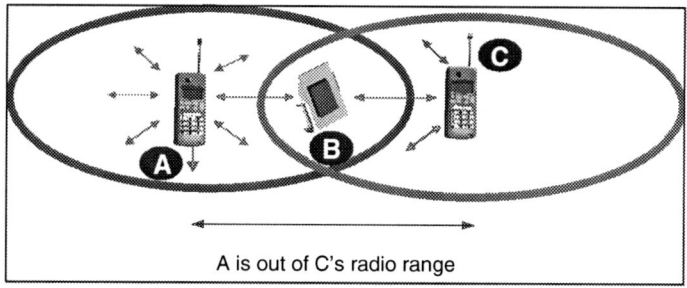

Fig. 2.25. Hidden stations

To understand the principle of CSMA/CA mechanism, we first of all examine the radio problems which brought to develop a new mechanism inspired of the CSMA/CD but adapted to the context of the wireless transmission. Indeed, the wireless transmission raises unknown new problems such as:

- The hidden stations problem

- The exposed stations problem

- The near-far effect related to radio signal attenuation according to the covered distance

Hidden stations problem. The deployment of WLANs can face the hidden nodes problem (Figure 2.25). This problem occurs when two stations cannot hear each other owing to the fact that the distance which separates them is too large or that an obstacle prevents them from communicating between each other. But they have overlapped covering zones. If stations A and C make only the detection of the carrier by listening the channel, not being capable of hearing each other, they will be authorized to emit at the same time. Consequently, if A and C wish to send packets at the same time to station B located in the intersection of the covering zones, a collision occurs and thus B will not be able to receive any of the two data communications. Stations A and C are said to be hidden one from each other.

Exposed stations problem. Let us consider the case where station B transmits data to station A (as shown in Figure 2.26). If station C senses the radio channel, it can hear a communication in progress. It concludes that it cannot transmit packets to station D but if C transmits, collisions would be created only in the area between B and C and not in the area where destinations D and A are.

In order to overcome these problems, the IEEE 802.11 standard uses the collision avoidance mechanism with a positive acknowledgement scheme as follows. A station wanting to transmit senses the medium. If the medium is

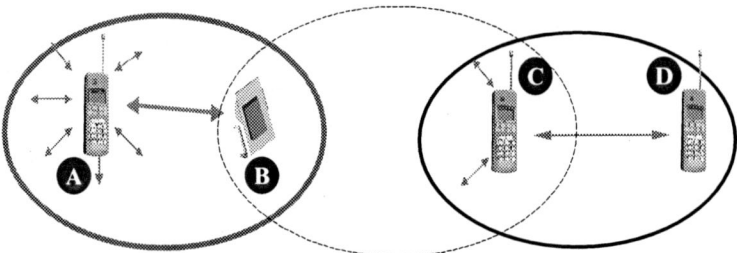

Fig. 2.26. Exposed stations

busy, the transmission is differed. If the medium is free for a specific time (called distributed interframe space [DIFS], in the case of the asynchronous data transmission), then the station is authorized to transmit. The receiving station (which can be also the AP playing its relay role) checks the cyclic redundancy check (CRC) of the received packet and returns an acknowledgement (ACK) of delivery. The reception of the ACK indicates to the transmitter that no collision took place. If the transmitter does not receive the acknowledgement, then it retransmits the frame after an ACK_TIMEOUT until it obtains it or gives up after a given number of retransmissions.

This kind of protocol, inspired from that defined in the Ethernet network, is effective when the medium is not overloaded, since it authorizes the stations to transmit with a minimum delay, but collisions can always take place. This is due to the fact that the stations which sense the medium, can find it free, and finally decide to transmit, sometimes at the same time as other stations carrying out this same series of operations.

These collisions must be detected, so that the MAC layer can retransmit the frame without having to pass again by the high layers, what would generate significant delays. Indeed, a first difference between CSMA/CD and CSMA/CA mechanisms is the possibility of collisions detection. In CSMA/CD, collision is detected at transmission side because the stations are capable of sensing the medium during data transmissions. But the collision can be detected only at the receiver side in the case of the CSMA/CA mechanism. In the Ethernet networks the collisions are detected by the transmitting stations which retransmit data frames using a recovery algorithm called binary exponential back-off (BEB).

In order to supervise the network activity, the MAC sublayer works in collaboration with the physical layer. The physical layer uses clear channel assessment (CCA) algorithm to evaluate the availability of the channel. To know if the channel is free, the physical layer measures the power received by antenna called received signal strength indicator (RSSI). The physical layer thus determines that the channel is free by comparing the RSSI value with a fixed threshold and transmits thereafter to the MAC layer an indicator of free channel. In the contrary case, the transmission is differed according to rules' quoted previously.

All in all, the CSMA/CA protocol is based on:

– Sensing the medium thanks to CS procedure (CS carrier sense)

– Using interframe space (IFS) timers

– Using positive acknowledgements and the collision avoidance approach

– Executing backoff algorithm

– Using multiple access

Interframe Spaces (Spaces Between Successive Frames). IEEE 802.11 standard defines four types of IFS timers classified by ascending order, which are used to define different priorities:

- Short interframe space (SIFS) is used to separate the transmissions belonging to the same dialogue (data frames and acknowledgements). It is the smallest gap between two frames. There is always, at most, only one station authorized to transmit at any given time, taking thus priority over all other stations. This value is fixed by the physical layer and is calculated in such way that the transmitting station will be able to switch back to receive mode to be able to decode the incoming packet. A high priority SIFS is then used to transmit frames like ACK, CTS and response to a polling.

- Point coordination IFS (PIFS) is used by the AP (called coordinator in this case) to gain the access (to the medium before any other station. It reflects an average priority to transmit the time-bounded traffic.

- Distributed IFS (DIFS) is the IFS of weaker priority than the two previous; it is used in the case of data asynchronous transmission.

- Extended IFS (EIFS) is the longest IFS. It is used by a station receiving a packet which is corrupted by collisions to wait more time than the usual DIFS in order to avoid future collisions.

These timers define degrees of priority. When several stations wish to transmit simultaneously, the station wishing to emit the urgent frames as the acknowledgements will be able to send them in first. Then other priority frames will be transmitted considered to be like those related to the administration tasks or the traffic which has delay constraints. Lastly, the least important information concerning the asynchronous traffic will be transmitted after a longer latency.

Table 2.5. *Main parameters*

Parameter	Physical layer	Value
Slot Time	FHSS	50
	DSSS, HR/DSSS	20
	OFDM	9
	IR	8
SIFS Time	FHSS	28 microsec $+/-$ 10%
	DSSS, HR/DSSS	10 microseconds
	OFDM	16 microseconds
	IR	10 microseconds
CW_{min} (slots)	FHSS	15
	DSSS, HR/DSSS	31
	OFDM	15
	IR	63
CW_{max} (slots)	FHSS/DSSS/IR/OFDM	1023

PIFS and DIFS are calculated (formulae [2.1] and [2.2]) according to SIFS.

$$\text{PIFS} = \text{SIFS} + \text{SlotTime} \qquad (2.1)$$

$$\text{DIFS} = \text{SIFS} + 2 \times \text{SlotTime} \qquad (2.2)$$

where:

SlotTime = minimal delay to determine the state of the channel + Rx − Tx turnaround time + propagation time.

The value of SIFS is related to the characteristics of the physical layer (see Table 2.5).

Note that in the IEEE 802.11e, we have an EDCF where these timers take each two possible values: one low and one high. By this way, we make a natural differentiation of traffic over the link.

Another Note concerns slow traffic; as we can see if we mix a station sending on the DCF with a speed of 1 Mbps with another using MIMO at 190 Mbps the slow station will monopolize the channel 190 times more than the rapid station. The only way to avoid that is to configure APs where low speeds are disabled.

Exponential Backoff Algorithm. This algorithm is a well-known method used to resolve the problem of equity of competitor accesses of several stations to a shared medium. This method requires that each station chooses a "withdrawal period" a random number N (called backoff time) ranging between 0 and the size of contention window. This number N is expressed in slots. Before accessing to the medium, each station must check that another station did not access the medium before it. It waits for this number of slots before accessing the medium.

The duration of a slot (SlotTime) is defined so that the station is always able to determine if another station has accessed the medium at the beginning of the preceding slot. The average value of the waiting delay grows exponentially with the number of retransmissions.

The IEEE 802.11 standard defines the backoff algorithm to be executed in the following cases:

- When a station wanting to transmit, senses the medium before the first transmission of a frame and finds that the medium is occupied, it waits during a random period before trying to transmit again
  - After each retransmission
  - After a successful transmission

The only case where this mechanism is not used is when the station decides to transmit a new frame and the medium has been free for more than SIFS/PIFS/DIFS.

The waiting random duration (WRD) is calculated by using the following formula:

$$\text{WRD} = \text{CW}^*\text{random}()^*\text{SlotTime} \quad (2.3)$$

where:

- CW is the size of the contention window, $\text{CW} = [\text{CW}_{min} - \text{CW}_{max}]$. At each failure, the CW size is doubled until reaching a maximum value CW_{max}. The values of CW_{min} and CW_{max} are given in Table 2.5.
- Random is a random number ranging between 0 and 1.
- SlotTime = minimal delay to determine the state of the channel +Rx − Tx turnaround time + propagation time.

Reduction in the withdrawal period in the case of CSMA/CA mechanism is different from that applied in the case of CSMA/CD procedure. Indeed, instead of a reduction at each clock signal, the number of slots is fallen only when the channel is free throughout this slot. If the channel becomes busy, in this case the timer is frozen. It is an additional protection in order to decrease the probability of collisions after the random waiting duration.

We assume that the stations have the same probability of accessing the medium because each station must reuse the same algorithm after each retransmission. The disadvantage of this method is that it does not guarantee a maximum delay and consequently is not appropriate for real-time applications (such as the voice and video).

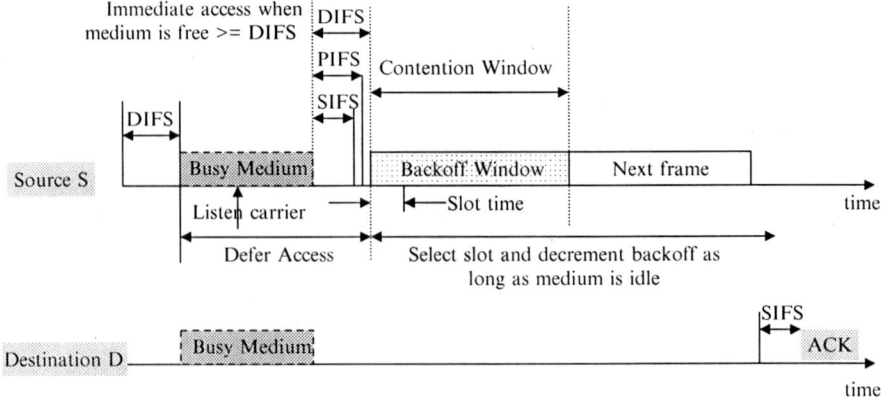

Fig. 2.27. CSMA/CA access method

Medium Listening. This function is fulfilled thanks to two procedures:

- A procedure for listening at the physical layer called physical carrier sense (PCS). PCS detects the presence of other stations by analysing all the frames on the wireless medium and by detecting the activity on the medium thanks to the relative energy of its signal compared to the other stations

- A procedure for listening at the MAC layer named virtual sense carrier (VCS) in the case of a reservation of the medium.

The basic global mechanism is described in Figure 2.27 by considering the case of the asynchronous data transmission between a source S and a destination D. A similar operation is adopted in the case of transmissions of other frames; we should only consider the adequate IFS.

5.3 Virtual Carrier Sense CSMA/CA Mechanism with Short Messages RTS and CTS

Because of collisions, packet loss can occur despite of using BEB recovery algorithm and acknowledgements of the frames. Consequently, in the case of an environment where there are a lot of hidden stations or in the case of transmissions of very long frames, the standard defines a virtual carrier sense mechanism. The principle of operation of the mechanism is described as follows.

A station wanting to send frames, begins by initially transmitting a short control packet called request to send (RTS), which contains the source, the destination and duration of the transmission (i.e. total duration of the transmission of the packet and its ACK). The destination station answers (if the medium is free) with a control packet called clear to send (CTS), containing the same duration information.

All stations receiving either the frame containing the request for reservation of channel RTS, or the frame of response of reservation CTS, set their virtual carrier sense indicator called network allocation vector (NAV), for the given duration, and use this information together with the PCS when sensing the medium. They consider that the support is busy for the duration period specified in the control packets RTS and CTS. If the source does not receive the CTS packet, it assumes that a collision occurred and will retransmit the RTS packet after a random waiting period according to the principle, described in paragraph 5.2. If the recipient receives the CTS packet correctly, the source emits an acknowledgement to announce to the recipient that the packet CTS was received. Then the communication will be able to take place.

This mechanism reduces the probability of collision which can be caused by a hidden station of the transmitter in the zone of the receiver thanks to the short duration of transmission of the RTS, because the station will hear the control packet CTS and will consider that the medium is like occupied until the end of the transmission. The Duration information in the RTS packet protects the transmitter area from the collisions during the transmission of acknowledgement (these collisions could be caused by the stations which are out of the radio range of the receiver station). Figure 2.28 shows a transmission between two stations A and B, and the NAV setting of their neighbours.

It should also be noted that thanks to the fact that RTS and CTS are short frames, the number of collisions is reduced and the overhead of collisions is reduced since these frames are recognized more quickly than if all the whole

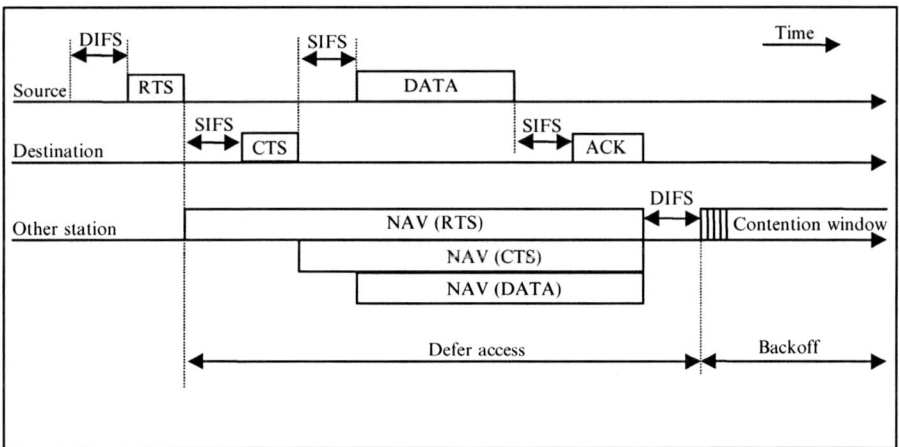

Fig. 2.28. CSMA/CA with RTS/CTS

MAC frame is transmitted (this is true if the packet is significantly bigger than the RTS, thus the standard authorizes the transmission of short packets without the exchange of RTS/CTS).

The mechanism CSMA/CA with RTS/CTS is a mechanism which thus allows the data transmission and the reception of the ACK without collision. Consequently, it is advised to also use it to send long frames for which a retransmission would be too expensive in terms of bandwidth. This is controlled per each station by a parameter called RTS_Threshold.

How can we solve the problem of hidden stations thanks to this mechanism? The hidden station problem is illustrated in Figure 2.25.

Station C is hidden station from station A. Let us consider the case where station A transmits data to the station B. Station C does not detect the activity of station A. In this case, station B can transmit freely without interfering with station C transmission. If A and B exchange RTS/CTS messages, station C is informed thanks to the sending of a frame CTS by station B indicating that the medium is busy. C thus does not try to transmit during the transmission between A and B.

If the virtual carrier sense is executed, the sending of control packet CTS by the recipient B can solve the problem of the hidden nodes. Indeed, thanks to its packet CTS, B will reach all the stations in its radio range and thus will inform all the stations which would like potentially to emit. If B receives the RTS packet of A in first, then B will receive the CTS packet emitted by B and thus will understand that the medium is busy and thus stops transmitting.

The IEEE 802.11 standard specifies that the basic technique DCF is mandatory and other procedures CSMA/CA with RTS/CTS and HCF are optional.

Finally, if the IEEE 802.11 cell is not crowded, simply avoid using RTS/CTS because you will loose a lot of time because of signalling.

5.4   PCF and HCF

Beyond the basic DCF, there was an optional PCF mechanism now replaced by IEEE 802.11E. This mechanism makes use of the higher priority that the AP may gain by the use of a smaller PIFS. PCF uses a centralized polling scheme, which requires the AP as a point coordinator (PC) in a BSS. The AP sends polling requests to stations for data transmission, hence controlling medium access. If PCF is supported, both PCF and DCF coexist and the time is divided into superframes. A superframe starts with a beacon frame. Each superframe consists of a CP where DCF is used and a CFP where PCF is used.

During the CFP, the PC maintains a list of registered STAs and polls each STA according to its list. Then, when an STA is polled, its gets the permission to transmit data frame. Since every STA is permitted a maximum length of frame to transmit, the maximum CFP duration for all the STAs can be known and decided by the PC, which is called CFP_max_duration. The time used by the PC to generate beacon frames is called target beacon transmission time (TBTT). In the beacon, the PC denotes the next TBTT and broadcast it to all the others in the BSS. In order to ensure that no DCF STAs are able to interrupt the operation of the PCF, a PC waits for a PIFS, which is shorter than DIFS, to start the PCF. Then, all the others STAs set their NAVs to the values of CFP_max_duration time, or the remaining duration of CFP in case of delayed beacon. During the CP, the DCF scheme is used, and the beacon interval must allow at least one DCF data frame to be transmitted. Normally, PCF uses a round-robin scheduler to poll each STA sequentially in the order of polling list, but priority-based polling mechanisms can also be used if different QoS levels are requested by different STAs.

Performance analysis of this protocol showed that it is not adapted to the transmissions of traffic with strong time constraints and thus there is no product on the market that implements it. It is possible that if a station takes the channel, it can monopolize it during a long time and the beacons frames can be delayed. HCF mechanism (IEEE 802.11e) deals with these aspects.

5.5 Frames Formats

There are three main types of frames:

- Data frames which are used for data transmission

- Control frames which are used to control access to the medium (e.g. RTS, CTS, ACK)

- Management frames which are frames transmitted to exchange management information but are not forwarded to upper layers (e.g. beacon frames)

Each frame type is subdivided into different subtypes according to their specific functions.

MAC Frame Format. The structure of the frame is traditional (close to that of an Ethernet frame) (Figure 2.29). For each IEEE 802.11 frame an associated MAC header is defined. The functionalities associated with these types of frames are described in section 6.

The following figure shows the general format of the MAC frame, some fields are only present in specific frames, as described later. The fields are as follows.

Frame control field. The structure of the frame control field is described below (Figure 2.30):

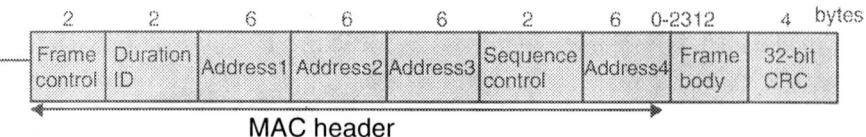

Fig. 2.29. MAC frame format

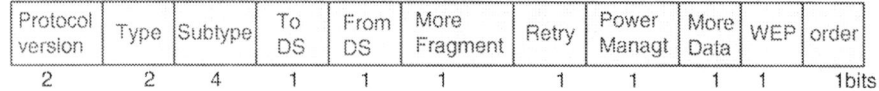

Fig. 2.30. Frame control field

This control field contains the following information:

- Protocol version: this field contains two bits which could be used to recognize possible future versions of IEEE 802.11 standard. In the current version of the standard, the value is fixed to 0.

- Type and sub-type: the six bits define the type and the sub-type of the frames (see Annex A).

- Towards the system of distribution (ToDS): this bit is set to 1 when the frame is addressed to the AP so that it forwards it to the DS. This concerns the case when the source and the recipient are in the same BSS and the AP must relay the frame. The bit is set to 0 in all other frames.

- FromDS (coming from the system of distribution): this bit is set to 1 when the frame is received from the DS.

- More fragments (of other fragments): this bit is set to 1 when there are more fragments following the current fragment.

- Retry (retransmission): this bit indicates that the fragment is a retransmission of a previously transmitted fragment. This is used by the receiver station to recognize duplicated transmissions of frames that may occur when an acknowledgement packet is lost or damaged.

- Power management (energy saving): this bit is set to 1 to indicate that the station will be in energy-saving mode (sleep mode) after the transmission of this frame. This is used by stations which are changing state either from power save to active or vice versa.

- More data (of other data): this bit is also used for the energy management. It is used by the AP to indicate that more frames are buffered for this station. The station may decide to use this information to require continuing polling frames or even changing to active mode.

- WEP (security): this bit set to 1 indicates that the body of the frame is encrypted according to the WEP algorithm.

- Order (order): this bit indicates that this frame is sent by using the strictly ordered service class. This class is defined for the users who cannot accept change of order between the unicast and multicast frames. The strictly ordered service class is defined for users who cannot accept change of ordering between unicast frames and multicast frames.

Duration/ID. This field has several uses depending on the frame type. It can be used as a duration field indicating the time (in microseconds) the channel will be allocated for successful transmission of an MAC frame. In some control frames, this field contains an association or connection identifier. In polling frames with energy-saving mode it contains the identifier ID of the station.

Address fields. A frame may contain up to four addresses, depending on the bits ToDS and FromDS defined in the control field (see paragraph 6.1) as follows:

- Address 1 always the recipient address (i.e. the BSS station that is the immediate recipient of the packet). If ToDS is set, this is the AP address, if ToDS is not set then this is the address of the end station.

- Address 2 is always the transmitter address (i.e. the station which is physically transmitting the packet). If FromDS is set, this is the AP address, if it is not set then it is the station address.

- Address 3 is in most cases the remaining, missing address. In a frame with FromDS set to 1, address 3 is the original source address, if the frame has the ToDS set, then address 3 is the destination address.

- Address 4 is used in special cases where a wireless DS is used, and the frame is being transmitted from one AP to another. In such cases, both the ToDs and FromDS bits are set, so both the original destination and the original source addresses are missing.

The transmitter address and receiver address are the MAC addresses of stations within the BSS and which are transmitting and/or receiving frames over the WLAN. The source address and destination address are the MAC addresses of the stations (wireless or not) which are the ultimate source and destination of the frame.

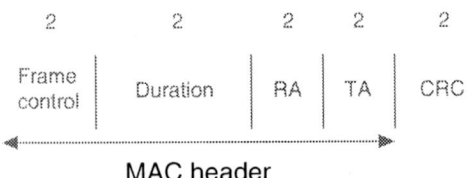

Fig. 2.31. RTS frame format

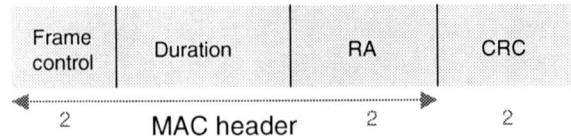

Fig. 2.32. CTS frame format

Sequence control field. The sequence control field is used to represent the order of the various fragments belonging to the same frame, and to recognize duplicated packets. It consists of two subfields, the number of fragment and the sequence number which define the number of frame and the number of the fragment in the frame sent between a given transmitter and receiver.

Cyclic redundancy check field. The error checking code is a 32-bit field formed starting from the following generating polynomial:

$$G(x) = x^{32}+x^{26}+x^{23}+x^{22}+x^{16}+x^{12}+x^{11}+x^{10}+x^{8}+x^{7}+x^{5}+x^{4}+x^{2}+x+1$$

Format of RTS, CTS and ACK Frames.

RTS frame format. The RTS frame (Figure 2.31) consists of the following fields:

- RA: the address of the receiver STA of the next data or management frame
- TA: the address of the station which transmits the RTS frame
- Duration: time, in microseconds, necessary to the transmission of the frame, plus one CTS frame, plus one ACK frame, plus three SIFS intervals

CTS format. The CTS frame (Figure 2.32) consists of the following fields:

- RA: the receiver address of the CTS frame, directly copied from the field transmitter address (TA) of the immediately previous RTS frame to which the CTS is a response.

Fig. 2.33. ACK frame format

Table 2.6. IEEE 802.11b throughputs and parameters

Bandwidth Mbps	Time intervals								Throughput Mbps
		Data frames				ACK frames			
	DIFS (µs)	Preamble (µs)	Overhead (µs)	Payload (µs)	SIFS (µs)	Preamble (µs)	Overhead (µs)	Total (µs)	
1	50	192	272	12,000	10	192	10	12,726	0.94
2	50	192	136	6,000	10	192	10	6,590	1.84
5.5	50	192	49	2,182	10	192	10	2,685	4.47
11	50	192	25	1,091	10	192	10	1,570	7.64

- Duration: the value obtained from RTS frame, minus the time of transmission, in microseconds, required to transmit the CTS frame and its SIFS interval.

ACK frame format. The ACK frame (Figure 2.33) looks as follows:

- RA: the field directly copied from the address 2 field of the frame immediately preceding this ACK frame.

- Duration: if the bit more fragment is set to 0 in the frame control field of the previous frame, the duration value is set to 0, otherwise it is the last value of the previous frame, minus time, in microseconds, required to transmit the ACK frame and its SIFS interval.

Overhead of IEEE 802.11 MAC Frames. The four theoretical throughputs suggested by the IEEE 802.11b standard are not really reached because of the overhead in the header and in the CSMA/CA. Table 2.6 summarizes the effective throughputs really offered.

Comparison Between MAC 802.11x and MAC 802.3 Frames. A comparison in terms of bandwidth and overhead between various technologies IEEE 802.11 and 802.3 is summarized in Table 2.7.

In addition, throughputs offered by IEEE 802.11a (without RTS/CTS) with packets of 1500 bytes length are listed in Table 2.8.

Notice that in practice these throughputs can be lower. The results are obtained with the IPERF tool sending TCP data chuncks.

Table 2.7. Comparison table between IEEE 802.11 and 802.3 standards

IEEE standard	802.11	802.11b	802.11a	802.3
Bandwidth	1,2 Mbps	1, 2, 5.5, 11 Mbps	6, 9, 12, 18, 24, 36, 48, 54 Mbps	10, 100, 1,000 Mbps
DIFS	128 µs	50 µs	6 µs	IFS 6.9 µs, 0.96 µs
Slot time	50 µs	20 µs	6 µs	5.12 µs, 5.12 µs
DATA preamble	250 µs	192 µs	10, 7, 6, 5, 4, 3, 3, 3 µs	6.4 µs, 0.64 µs
MAC overhead	34 bytes	34 bytes	34 bytes	18 bytes
Payload	46–1,500 bytes (0–2,312 w/o 802.3)	46–1,500 bytes (0–2,312 w/o 802.3)	46–1,500 bytes (0–2,312 w/o 802.3)	46–1,500
SIFS	28 µs	10 µs	13 µs	–
ACK preamble	192 µs	192 µs	10, 7, 6, 5, 4, 3, 3, 3 µs	–
MAC overhead/ACK	14 bytes	14 bytes	14 bytes	–

Table 2.8. IEEE 802.11a rates

Theoretical throughput (Mbps)	Effective throughput (Mbps)
6	5.3
24–27	18
36	24
54	32

6. Functions

Several functions are supported by the IEEE 802.11 standard. This section aims to describe in a brief way the characteristics of each of them.

6.1 Addressing

A compliant Ethernet addressing on 48 bits is adopted. The standard defined four fields which are:

- Address 1 is always the address of the receiver (i.e. the station in the BSS which is the receiver of the packet). If ToDS is set to 1, it is the address of the AP, if not, it is the address of the station.

- Address 2 is always the address of the transmitter (i.e. the station that physically transmits the packet). If FromDS is set to 1, it is the address of the AP, if not, it is the address of the transmitting station.

Table 2.9. IEEE 802.11 addresses

scenario	to DS	from DS	address 1	address 2	address 3	address 4
ad-hoc network	0	0	DA	SA	BSSID	-
infrastructure network, from AP	0	1	DA	BSSID	SA	-
infrastructure network, to AP	1	0	BSSID	SA	DA	-
infrastructure network, within DS	1	1	RA	TA	DA	SA

DS: Distribution System
AP: Access Point
DA: Destination Address
SA: Source Address
BSSID: Basic Service Set Identifier
RA: Receiver Address
TA: Transmitter Address

- Address 3 is the address of the original transmitter when the FromDS field is set to 1. If not, and if ToDS is set to 1, address 3 is the destination address.

- Address 4 is used in a special case, when the DS is used and a frame is transmitted from an AP to another. In this case, ToDS and FromDS are both set to 1 and it is thus necessary to inform at the same time the original transmitter and the recipient.

Table 2.9 summarizes the use of the various addresses according to FromDS and ToDS bits.

6.2 Association, Reassociation and Disassociation

An IEEE 802.11 station cannot send a message via an AP (or another station in the case of an ad hoc configuration) only if it executes an association beforehand. The association procedure is always launched by the station. For simplicity, let us consider the case of an architecture with APs. When a station wants to join a BSS or an ESS, it must choose an AP with which it joins according to some criteria: signal power, signal quality and network load. A station can be associated only with one AP. The AP assigns an association identifier called AID (association ID) for each associated station. The AID is a 16-bit integer. The AP informs the DS of the association. The association process consists of an exchange of information relative to the various stations, capacities of the cell, etc. DS records station identifier, its AID and its current position. Any station, being in the area managed by the BSS, can thus transmit or receive data frames only after the end of the association which must finish successfully. The association procedure is preceded or combined with an authentification phase. This is shown in some APs that by "association pending" field until the authentication takes place.

Disassociation can be carried out either at the request of the AP or the station which launched the association operation. Before the association procedure, a station needs to get synchronization information from the AP or from the other stations in an ad hoc configuration. Disassociation is a well known attack when security is disabled and targets in general identity usurpation of the station.

Concerning the reassociation operation, it takes place in several cases which are:

- Roaming

- Change of the characteristics of the radio environment

- Traffic network too high on the AP

- Load-balancing activated

To be able to join a BSS/IBSS, an IEEE 802.11 station needs to *ranging* or listening mechanisms. It has to detect the presence of an AP, respectively another station in the case of an ad hoc configuration. The station can get the needed information by one of two means: passive scanning and active scanning.

Passive and Active Scanning. Similarly to other wireless and mobiles systems, any IEEE 802.11 station, scans the environment, i.e. it senses the medium to intercept the beacons signals (a beacon is the frame sent periodically (containing the synchronization information) by the AP, equivalent to the broadcast control channel used in GSM networks). If one of the selection criteria of the target AP is not respected (for example, power of transmission of the AP is weak) the station continues its search to find a suitable AP. This search is done in two manners: passive or active listening. These procedures are based on performances or energy consumption criteria.

Passive scanning: in this case, the station just waits to receive a beacon frame coming from the AP.

Active scanning: in this case, once the station found the appropriate channel, it directly sends a request of association through a probe request frame and waits until the AP answers with a probe response frame. Once the operation is successfully completed, the station is synchronized on the frequency of the AP. Periodically, the station scans all the channels to evaluate if another AP has better performances and in this case joins with it at the condition that the SSID is allowed.

In this case, the station activates the active listening and sends registration requests to the various APs which are in its radio zone. Upon receiving beacons from the APs, it evaluates the received signals and chooses the more adapted AP according to its selection criteria. An authentication phase thus takes place to be able to register the station. Once registered in the BSS/ESS, the station sends an association request. The station will not be able to start a transmission phase unless the AP answers positively to the formulated association request.

The procedure of reassociation is described in paragraph 6.4.

6.3 Fragmentation and Reassembling

Typically, fragmentation is used in wireless communications to optimize the reliability of transmission. Longer frames are thus divided into small fragments.

More particularly, in a WLAN environment, it is necessary to handle packets of small size for the following reasons:

1. The error rate per bit is more important on a radio connection than on a fixed physical medium. The probability of a packet of being corrupted increases proportionally with the packet size.

2. The overhead generated by the successive retransmissions of corrupted packets (because of collisions or even because of the noise): more the sizes of the packets are small, while having a minimal size for which the overhead remains negligible compared to the payload of the frame, less this overhead is important.

3. On the other hand, more frames induce more timers and thus longer delays. So it is always a matter of compromises between losses, delays and throughput.

An IEEE 802.11 frame must be fragmented if its size exceeds a threshold value called Fragmentation_Threshold set by the network administrator. When a frame is split up, all the fragments are transmitted in a sequential way. The medium is not released until all the fragments are successfully transmitted. If an ACK is not correctly received the station stops transmitting and tries to reach the medium again and takes again the transmission starting from the last not delivered fragment. If the stations use RTS/CTS mechanism, only the first fragment sent uses RTS/CTS frames as it is illustrated in Figure 2.34; the channel will thus be reserved for the total duration of the transmission of fragments. The transmission of each fragment is framed by two SIFS and the stations unconcerned by the exchange update gradually their NAVs respectively.

At the receiving side, a procedure of reassembly of fragments is needed.

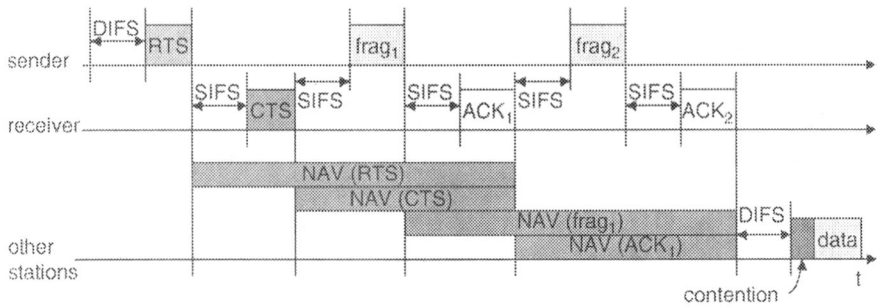

Fig. 2.34. CSMA/CA with RTS/CTS – fragmentation/reassembling

Procedures of fragmentation/reassembly are rather simple; they are based on an algorithm sending and waiting for acknowledgements, where the transmitting station is not authorized to transmit a new fragment as long as one of the two following events did not occur:

- Reception of an ACK of the fragment
- Decision that the fragment was retransmitted too many times and drops the whole frame

It should be noted that the standard authorizes the station to transmit to a different address between retransmissions of a given fragment. This is particularly useful when an AP has several outstanding packets to several different destinations and that one of them does not respond. Fragmentation is possible only with frames with a unicast destination address.

In general, we should never promote fragmentation in the network but rather do it at the application level.

6.4 Roaming

The principle of roaming is equivalent to the handover in cellular architectures like GSM, but which takes place on the layer-2 level. The standard does not give details on the implementation of this functionality but defines the basic rules. Those include active/passive listening, synchronization procedure, association and reassociation procedures. The procedures are now enriched with the necessary pre-authentication architecture that enables the station to send a request to a remote AP asking to derive the necessary keys before a handover takes place. This is dealt in the IEEE 802.11R standard.

Roaming allows a user to freely move from a cell to another without losing the access to the network. This function is similar to the handover, with a major difference; on a LAN, during packet transmission, the transition from a cell to another must be made between two transmissions of packets, contrary to mobile networks where the transition can occur during a communication. This makes the roaming relatively easier in the LAN. The nature of the bridged architectures is that once switches have noticed that a MAC address left a port and is now on a different port they automatically route the frames to the new place. This procedure can be very quick.

In the case of voice traffic, a temporary disconnection may affect the conversation, whereas in data transmission, the performances could be considerably reduced because of the retransmission which will be carried out by the protocols of the higher layers.

During its displacements, an IEEE 802.11 station may change its connection with one or more APs. It will have to choose among them to join according to some fixed criteria (power of the signal, quality of the transmission, etc.). Once a suitable transition candidate has been found, the STA leaves its current AP and associates with the target AP. Once accepted by the new AP, the station is frequency-synchronized on the radio channel of this one. Periodically, the associated station scans all IEEE 802.11 channels to determine if another AP is likely to offer better performances. If it is the case, it associates with the new AP, and senses the radio channel. In a similar way, this station carries out an operation of reassociation.

A reassociation, after a roaming phase, occurs in general when the station moved away from the initial AP, involving consequently a weakening of the signal. It can also be carried out because of a change in the radio characteristics or because of the increase in the traffic load on the initial AP. In this last case, the function is used for balancing the load, since it distributes more effectively the total load of the WLAN on the available wireless infrastructure.

This process of dynamic association/reassociation thus allows the administrator of the network to create a very wide cover by overlapping multiple IEEE 802.11 cells on the whole of the zone to cover (building, campus). With this intention, it can use the function of "reuse of the channels" when trying to configure each AP on DSSS 802.11 channels different from those used by the contiguous APs. Knowing that in the case of the use of the technique DSSS, we have the possibility to use 14 channels with partial covering from which only three of them are completely disjoint. These three channels are adapted to a multicellular coverage.

The process of roaming provides several advantages that are:

- Ease of coverage extension

- Network load-balancing

- Scalability of the network

- Transparency for the users

Discovery. The STA may develop a list of potential transition candidates so that it can make a transition to a target AP as quickly as possible at the appropriate time. Fast BSS transition services provide mechanisms for the STA to communicate and retrieve information on candidate APs prior to making a transition. The communications with the candidate AP can take place by direct communication or through the STA's existing association with its current AP.

An advanced procedure has been defined in 802.11 R for fast roaming. In this case, authentication happens with new candidate APs through associate ones.

6.5 Synchronization

When users move, their terminals must remain synchronized to be able to communicate. Within a BSS, the stations synchronize their clocks with the clock of the AP. To maintain a total synchronization within a cell, the AP periodically sends beacons frames which contain the value of the clock of the AP. At reception of these frames, the stations update their clocks. The structure which maintains the timers of all the stations of the same cell synchronized is called timing synchronization function (TSF); it is periodically broadcasted in the beacon frame.

The AP sends periodically beacons which contain a copy of TSF. These beacons can be delayed because of the activity which can take place on the medium. If the local timer TSF of a station is different from the stamp in the received beacon, the station updates its timer at the value of the received stamp. A similar procedure is carried out in the case of an ad hoc architecture. A more complex mechanism of distributed synchronization is performed to synchronize stations in the same IBSS. Each station in an IBSS adopts the received synchronization of any beacon which has a value TSF more recent than its own timer TSF. Moreover, each station calculates an additional random period to send the beacon (with TSF included) on the medium. TSF is based on an internal clock of 1 MHz with ticks in microseconds.

The structure TSF is given in Appendix B.

6.6 Energy Conserving

Because of limited-battery equipments, it was necessary to define an energy-saving mechanism. Two operating modes are defined: the active mode where the terminal constantly listens to the medium and the energy-saving mode where the terminal can be in a sleeping state, it is known as power save polling mode.

In the case of architecture with APs, the principle is as follows (see Figure 2.35). The AP maintains a database where it collects information concerning the stations in energy-saving mode. It stores also all the data frames which are addressed to them. A station prevents the AP before being in an idle state. Thus, the AP gathers its frames arrived during its sleeping period. The

Wi-Fi: Architecture and Functions

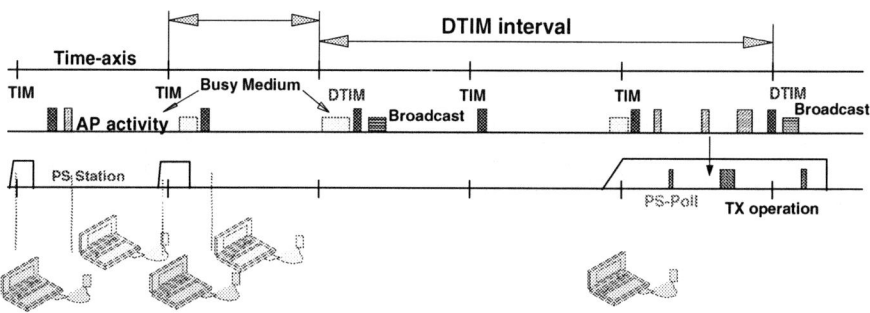

Fig. 2.35. Energy conserving (infrastructure mode)

stations awake periodically to consult a called specific frame traffic indication map (TIM) sent by the AP concerning the unicast traffic which it stores. If the AP gathers data frames for a specific station, this latter sends a request to the AP "Polling Request Frame" indicating that it will be able to receive its data. Between the moments of reception of TIM frames, the terminals turn over in the idle mode.

In order to distinguish between the broadcast/multicast traffic and the unicast traffic, the AP sends a called specific frame delivery traffic indication map (DTIM) and the same mechanism is applied. The period of the sendings of DTIM frames is usually taken as a multiple of the period of TIM frames' transmission. TIM and DTIM structures are given in Appendix B.

In Figure 2.35, the first axis shows the intervals of beacons. The second axis shows the activity of the AP. The AP sends periodically beacons but these latter can be delayed because of the activity on the wireless medium. The third axis shows the stations which awake periodically to listen beacons and TIM frames. If a station wants to receive data from the AP, it sends to it a request PS-poll (power save polling request frame). The field "Duration" in the PS-poll frame is replaced by the AID. If the bufferized data have an important size, then fragmentation is applied.

A last word on energy saving is controversial. In fact, we have two problems: AP buffers cannot be infinite and hence cannot store forever frames addressed to sleeping stations and second it is very easy to build a DoS (Denial of Service) attack on an AP by sending fake traffic to sleeping nodes (that can also be fake).

6.7 Management Frames Format

Functionalities of association/disassociation, authentication, synchronization and energy focus on the use of management frames which transport various structures which are summarized in Appendix B:

- Association, reassociation and disassociation
- Authentication, deauthentication

- Beacon frame
- Probe
- TIM and DTIM

7. Mobility

Mobility is a critical issue to solve since the context of IEEE 802.11 systems gives the possibility to the users to freely roam through several BSS, IBSS and ESS. Obviously, the user wishes to keep connectivity in a transparent way during his displacements. IEEE 802.11 offer a mechanism of roaming which makes it possible to manage this mobility on the level of layer 2. It is clear that the standardization of the protocol IAPP, by the working group IEEE 802.11F, contributes to facilitate the interworking of different networks. In the case of seamless roaming, several aspects must be taken into account:

- The attribution of IP addresses. The mobile user can belong to the same sub-net or distinct IP networks. A DHCP server will be generally used; some APs also support this functionality. One can also use private IP addresses.

- The management of mobility at layer 3 to maintain network connections when IP addresses change dynamically. A candidate protocol, currently existing, is Mobile IP (RFC 2002), although this latter is not very adapted to a wireless context and gives dramatic handover times.

- Authentication, access control to the network and access control to the services, in particular within the framework of an interconnection of heterogeneous networks operated by various operators including particularly hot spots; this can be solved with an adequate IEEE 802.11r solution.

- Billing and QoS aspects (bandwidth sharing control).

8. Security

The problems of security and privacy complicate the installation of a WLAN. Even if certain security mechanisms are natively integrated, their security can be easily violated if the relevant precautions are not taken into account. The IEEE 802.11 Committee defines a solution to security by working out at the beginning a protocol called WEP. Serious failures were quickly detected in the proposed mechanisms and overcame lately. Mainly, short keys and the worse key derivation algorithm that one can design were the reason for that failure.

The reader can find in chapter 6 an exhaustive presentation of the basic security mechanisms of IEEE 802.11 standard, faults and works in progress.

9. IEEE 802.11 Family and its Derivative Standards

An evolution in terms of throughput led to the specification of four main standards IEEE 802.11b, 802.11g, 802.11a and recently 802.11n. Major drawbacks of the basic WLAN technology have been addressed and largely resolved in terms of regulation, security and roaming. A panoply of tracks outcome with solutions within several working groups (IEEE 802.11e, 802.11F, 802.11g, 802.1h, 802.1i, 802.1X).

9.1 IEEE 802.11g

This standard operates in the 2.4 GHz band (like IEEE 802.11b) and offers similar throughputs to those proposed by IEEE 802.11a. The modulations used are: BPSK, QPSK, CCK and OFDM (12–54 Mbps). The IEEE 802.11g products are compatible with Wi-Fi products. A broader availability of these products is now effective. We can thus reach an aggregate throughput of 162 Mbps divided between three networks using three completely disjoint channels. The modulation scheme used in IEEE 802.11g is OFDM for the data rates of 6, 9, 12, 18, 24, 36, 48 and 54 Mbps, and reverts to (like the IEEE 802.11b standard) CCK for 5.5 and 11 Mbps and DBPSK/DQPSK + DSSS for 1 and 2 Mbps. Even though IEEE 802.11g operates in the same frequency band as IEEE 802.11b, it can achieve higher data rates because of its similarities to IEEE 802.11a. The maximum range of IEEE 802.11g devices is slightly greater then that of IEEE 802.11b devices, but the range in which a client can achieve full (54 Mbps) data rate speed is much shorter then that of IEEE 802.11b.

Most of the dual-band IEEE 802.11a/b products became dual-band/tri-mode, supporting a, b and g in a single mobile adaptor card or AP. Despite its major acceptance, IEEE 802.11g suffers from the same interference as IEEE 802.11b in the already crowded 2.4 GHz range. Devices operating in this range include microwave ovens, Bluetooth devices and cordless telephones.

Chip maker Atheros sells a proprietary channel linking feature called Super G for manufacturers of APs and client cards. This feature can boost network speeds up to 108 Mbps by using channel linking. Also range is increased to four times the range of IEEE 802.11g and 20 times the range of IEEE 802.11b. This feature may interfere with other networks and may not support all b and g client cards. In addition, packet bursting techniques are also available in some chipsets and products which will also considerably increase speeds. This feature may not be compatible with other equipments. Broadcom, another chipmaker, developed a competing proprietary frame bursting feature called "125 High Speed Mode" or Linksys "SpeedBooster", in response to criticism of Super G's interference potential.

9.2 IEEE 802.11e

It makes revision to the MAC layer of the current IEEE 802.11 to expand support for LAN applications that have QoS requirements. DCF/PCF mechanisms provided by IEEE 802.11 MAC to support best-effort services are replaced with a new super set of features. IEEE 802.11e introduces the HCF. HCF comprises two principal mechanisms:

- A new contention-based channel access method called EDCA; it is an extension of the legacy DCF mechanism which integrates priorities and defines four traffic classes.

- A contention-free channel access method called HCCA.

With the EDCA, QoS is supported by using four access categories, each corresponding to an individual prioritized data flow. With the HCCA, a hybrid coordinator (HC) allocates transmission opportunities (TXOPs) to wireless stations by polling, so as to allow them contention-free transfers of data, based on QoS policies.

One main feature of HCF is to introduce four access category (AC) queues and eight traffic stream (TS) queues at MAC layer. When a frame arrives at MAC layer, it is tagged with a traffic priority identifier (TID) according to its QoS requirements. The frames with TID values from 0 to 7 are mapped into four AC queues using EDCA access rule. On the other hand, frames with TID values from 8 to 15 are mapped into eight TS queues using HCCA rule. The reason of separating TS queues from AC queues is to support strict parameterized QoS at TS queues while prioritized QoS is supported at AC queues. Another main feature of the HCF is the concept of TXOP, which is the time interval permitted, for a particular station to transmit packets. During the TXOP, there can be a series of frames transmitted by a station separated by SIFS. The TXOP is called either EDCF-TXOP, when it is obtained by winning a successful EDCA contention; or polled-TXOP, when it is obtained by receiving a QoS CF-poll frame from the AP. The maximum value of TXOP is called TXOPLimit, which is determined by AP.

Target applications include transport of voice (VoWLAN), audio, video, video conferencing and streaming.

9.3 IEEE 802.11d

This working group studies the lawful aspects covering the totality of the continents. IEEE 802.11d is referred to as a regulatory domain update. It deals with issues related to regulatory differences in various countries.

9.4 IEEE 802.11F

The goal of this working group was to develop an IAPP that allows interoperability between APs of multiple vendors across a DS supporting IEEE 802.11 wireless links.

9.5 IEEE 802.11h

This standard deals with power and spectrum management issues. The goal is to enhance the current IEEE 802.11 MAC and 802.11a PHY specifications with network management and control extensions to provide spectrum and TPC management in the unlicensed ISM 5 GHz band. Procedures of dynamic frequency selection (DFS) and TPC are defined.

9.6 IEEE 802.11i and IEEE 802.1X

These two standards deal with security features applied to IEEE 802.11b/a/g/n. Details on protocols and architectures are given in chapter 6.

9.7 IEEE 802.11k

RRM is a key enabler to the next generation of WLANs. It is an amendment for RRM. WLAN radio measurements enable stations to observe and gather data on radio link performance and on the radio environment. A station may choose to make measurements locally, or may request a peer station within the BSS to make one or more measurements and returns the results. It provides knowledge about the radio environment to improve performance and reliability.

The existing IEEE 802.11 measurements and information about them are inadequate to move ahead to the next generation of WLAN. The proposed 11k approach is to add measurements that extend the capability, reliability and maintainability of WLANs through measurements and provide that information to upper layers in the communications stack. In addition, there are new technologies needing more measurement and capability. These technologies include voice over IP (VoIP), video over IP and location, as well as mitigation of harsh radio environment places where WLANs are operated (multifamily dwellings, airplanes, factories, municipalities, etc.).

The group has provided the mechanisms using request/response queries and the MIB. The 11k specification adopted the TGh layer management request/response model to gather the information. Also, by recognizing the mobility requirements of the new technologies, such as VoIP and video streaming, the group specified measurements, the channel load report and the neighbour report as a means of developing pre-handoff information gathering that drastically speeds up the ability to handoff between cells within the same

ESS. By accessing and using this information, the stations (either in the APs or in the individual devices) can make decisions about the most effective way to utilize the available spectrum, power and bandwidth for its desired communications.

The RRM specification:

- Enables better diagnostics of problems

- Enables better frequency planning, optimized network performance, simplified and/or automated WLAN radio configuration

- Achieves better performance in dense BSS deployments

- Better utilize radio resources across client stations

- Enables new services

    - Location-based services
    - VoIP
    - Video over IP

The specification makes lists of known stations (stations, both APs and devices) visible to the requester; it includes link statistics including counts of MSDU/MPDU and channel utilization in total microseconds. It includes data rates, modulation and link margin as seen by other stations. It includes optional events to notify management stations of authentication and association events and can provide TRAPS on association and authentication failures. It also provides reports on all the pre-authentication and deauthentication events, as well as normal association and disassociation events. We also find transmitter power control (TPC) report, AP channel, power capability, neighbour report element ID, TSF information, received channel power indicator, DS parameter set, channel number, regulatory class, measurement duration, BSSID, threshold/offset, measurement modes (passive/active or beacon table), measurement duration, PHY type, beacon interval, number of frames, transmitted fragment count, multicast transmitted frame count, failed count, retry count, frame duplicate count, ACK failure count, transmitted frame count, BSSID information, channel number, spectrum management capability, QoS capability, etc.

It is the first step in making WLAN smart and capable of making appropriate decisions for fast handoff, for mesh connectivity, and for managing the radio environment.

9.8 IEEE 802.11j

This standard studies the convergence of the American standards IEEE 802.11 and European Hiperlan 2 (both functioning in the band of 5 GHz). An adaptation to the Japanese legislation is added with the use of 4.9 and 5 GHz bands.

9.9 IEEE 802.11p

The standard IEEE 802.11p is also referred to as wireless access for the vehicular environment (WAVE). It defines enhancements to IEEE 802.11 required to support intelligent transportation systems (ITS) applications. This includes data exchange between high-speed vehicles and between these vehicles and the roadside infrastructure in the licensed ITS band of 5.9 GHz. The communications provided by WAVE generally occur over distances up to 1000 m between roadside stations and mostly high speed, but occasionally stopped and slow moving, vehicles or between high-speed vehicles. WAVE includes a number of new classes of applications that pertain to roadway safety (e.g. vehicle collision avoidance) and emergency services (e.g. those provided by police, fire departments, ambulances and rescue vehicles). These applications impact WAVE in a number of ways; most notably in the reliability of the data communications and the extremely low latencies required. Some critical applications require that the total time from first signal detection to completion of multiple frames of data exchange must be completed within the order of 100 ms. WAVE adds capability to simplify the operations and associated management in order to support fast access.

WAVE operations utilize a control channel and multiple service channels. Prioritized access in WAVE operations uses the EDCA mechanism.

The extension of the PHY for WAVE builds on the OFDM system. The WAVE radio frequency system occupies a licensed ITS radio services band, as regulated in the United States by the Code of Federal Regulations. Other regions and countries may allocate other bands in the 5–6 GHz range. The OFDM system provides WAVE with data payload communication capabilities of 3, 4.5, 6, 9, 12, 18, 24, and 27 Mbps in 10 MHz channels. The support of transmitting and receiving at data rates of 3, 6 and 12 Mbps is mandatory. WAVE has the option to operate on 20 MHz channels. If using the optional 20 MHz channel implementation, data payload capabilities of 6, 9, 12, 18, 24, 36, 48 and 54 Mbps can be supported. The support of transmitting and receiving at data rates of 6, 12 and 24 Mbps is mandatory for the optional 20 MHz configuration. In the context of this standard, "WAVE" refers to operation within the ITS band and not to operation in other bands.

Stations operating in WAVE shall be capable of transferring messages between the roadside and vehicles travelling at speeds up to 140 km/h with a packet error rate (PER) of less than 10% for PSDU lengths of 1000 bytes and between the roadside and vehicles at speeds up to a minimum of 200 km/h with a PER of less than 10% for PSDU lengths of 64 bytes. For vehicle-to-vehicle communications stations shall be capable of transferring messages at closing speeds of up to a minimum of 283 km/h with a PER of less than 10% for PSDU lengths of 200 bytes. Multipath and the effects of motion are addressed.

9.10 IEEE 802.11u

IEEE 802.11u is an amendment to the IEEE 802.11 standard to add features that improve interworking with external networks. IEEE 802.11 currently makes an assumption that a user is pre-authorized to use the network. IEEE 802.11u covers the cases where user is not pre-authorized. A network will be able to allow access based on the user's relationship with an external network (e.g. hot spot roaming agreements), or indicate that online enrolment is possible, or allow access to a strictly limited set of services such as emergency calls. From a user perspective, the aim is to improve the experience of a travelling user who turns on a laptop in a hotel many miles from home. Instead of being presented with a long list of largely meaningless SSIDs the user could be presented with a list of networks, the services they provide and the conditions under which the user could access them. The IEEE 802.11u proposal requirements specification contains requirements in the areas of enrolment, network selection, emergency call support, user traffic segmentation and service advertisement. The IEEE 802.11u standard is in its proposal evaluation steps.

9.11 IEEE 802.11v

IEEE 802.11v is the wireless network management standard for the IEEE 802.11 family of standards. TGV is working on an amendment to the IEEE 802.11 standard to allow configuration of client devices while connected to IEEE 802.11 networks. The standard may include cellular-like management paradigms. The IEEE 802.11v standard is still in its proposal stages.

9.12 IEEE 802.11r (Fast Roaming/Fast BSS Transition)

IEEE 802.11r is the unapproved IEEE 802.11 standard that specifies fast BSS transitions. This will permit connectivity aboard vehicles in motion, with fast handoffs from one AP to another managed in a seamless manner. Handoffs are supported under the "a", "b" and "g" implementations, but only for data (using IEEE 802.11F or IAPP). The handover delay is too long to support applications like voice and video.

The primary application currently envisioned for the IEEE 802.11r standard is VoIP, or Internet-based telephony) via mobile phones designed to work with wireless Internet networks, instead of (or in addition to) standard cellular networks. The delay that occurs during handoff cannot exceed about 50 ms (the interval that is detectable by the human ear). However, current roaming delays in IEEE 802.11 networks average in the hundreds of milliseconds. This can lead to transmission "hiccups," loss of connectivity and degradation of voice quality. Faster handoffs are essential for IEEE 802.11-based voice to become widely deployed.

Another problem with current IEEE 802.11 is that a mobile device cannot know if necessary QoS resources are available at a new AP until after a transition. Thus, it is not possible to know whether a transition will lead to satisfactory application performance.

IEEE 802.11r refines the transition process of a mobile client as it moves between APs. The protocol allows a wireless client to establish a security and/or QoS state at the target new AP which leads to minimal connectivity loss and application disruption. The overall changes to the protocol do not introduce any new security vulnerabilities. This preserves the behaviour of current legacy stations and APs. Fast BSS transitions can only take place within a mobility domain within an ESS.

Until that time, however, enterprises will need to use proprietary hardware from vendors such as SpectraLink to get fast roaming for applications such as VoIP. It should be noted that some wireless VoIP systems use the insecure WEP encryption to keep authentication time under the 20-ms threshold mark. More secure encryption types such as TKIP or AES-CCM (CCMP) involve handshakes that happen after association, and they can typically take 30–40 ms, potentially disrupting the call. IEEE 802.21 might have to use IEEE 802.11r facilities for roaming between IEEE 802.11 domains.

The amendment addresses solutions to two classes of network infrastructures from a QoS perspective: one where the transition-enabled AP is willing to provision QoS resources at (re)association time; and another where the AP needs to pre-reserve the network infrastructure resources before transitioning.

We can refer to IEEE 802.11e and IEEE 802.11k to get necessary information on the neigbourhood to assist in optimizing delays.

The fast BSS transition services are defined to:

- Improve the efficiency of radio channel use by combining resource allocation with authentication during connection establishment
- Enable the client station to perform key computations prior to reassociation, thereby minimizing a potential critical path computation and enabling an authenticated (re)association exchange
- Remove the four-way handshake race conditions described in chapter 6
- Enable allocation of resources at reassociation time

The fast BSS transition services include:

- A key management framework for security that allows the station and the AP to pre-compute keys
- A mechanism for negotiating the allocation of resources prior to or at (re)association

- A mechanism for exchanging the messages in the transition protocol either directly to the target AP ("over-the-air"), or by being relayed through the currently associated AP ("over-the-DS")

9.13   IEEE 802.11s

IEEE 802.11s is the unapproved IEEE 802.11 standard for ESS mesh networking. It specifies an extension to the IEEE 802.11 MAC to solve the interoperability problem by defining an architecture and protocol that support both broadcast/multicast and unicast delivery using radio-aware metrics over self-configuring multihop topologies.

The Standard is being defined by the IEEE Task Group (TGs), the aim of the project is to provide a protocol for auto-configuring paths between APs over – configuring multihop topologies in a wireless distribution system (WDS) to support both broadcast/multicast and unicast traffic in an ESS mesh using the four-address frame format or an extension.

Two main proposals Wi-Mesh and See-Mesh have merged to create the global single joint proposal.

Wi-Mesh Proposal.   The Wi-Mesh alliance (WiMA), which includes Accton Technology, ComNets, InterDigital Communications, NextHop Technologies, Nortel, Philips, MITRE, Naval Research Laboratory, Swisscom Innovations and Thomson, has presented a proposal that enables seamless communications for wireless users regardless of equipment vendor. Wi-Mesh proposal is designed to work for all three major applications of mesh technology – consumer and small business, metropolitan and military.

See-Mesh Proposal.   Another consortium, See-Mesh, is backed by Intel, Nokia, Motorola, NTT DoCoMo and Texas Instruments. As part of their IEEE 802.11s proposal, Intel has introduced what they call mesh portals. Mesh portals offer interoperability to mesh networks by allowing older (and newer) wireless standard technology to be recognized and incorporated into the network.

IEEE 802.11s defines a complete framework that describes in detail the main features of WLAN mesh architecture, it includes:

- Operational modes (lightweight mesh operation, power saving mode)

- A reference model for interworking

- Multichannel operation (single and multiple radio devices)

- Mesh topology discovery and formation (neighbour discovery, path selection, peer link setup and channel selection)

- Mesh path selection and forwarding
  - As default protocol hybrid wireless mesh protocol (HWMP)
  - Optional, radio aware OLSR
- Security framework
  - Centralized/distributed/pre-shared key 802.1X authentication model
  - Extension AKM
- Multichannel MAC
- Congestion control
- Configuration and management
- Beaconing and synchronization
- Power management

9.14   IEEE 802.11w

IEEE 802.11w is the protected management frames standard for the IEEE 802.11 family of standards. TGw is working on improving the IEEE 802.11 MAC layer to increase the security of management frames. WLANs send system management information in unprotected frames, which makes them vulnerable. IEEE 802.11u provides mechanisms that enable data integrity, data origin authenticity, replay protection and data confidentiality for selected IEEE 802.11 management frames including but not limited to action management frames, deauthentication and disassociation frames.

This standard will protect against network disruption caused by malicious systems that forge disassociation requests that appear to be sent by valid equipment. It is expected that IEEE 802.11w would extend IEEE 802.11i to apply to IEEE 802.11 management frames as well as data frames. These extensions will have interactions with IEEE 802.11r and IEEE 802.11u. The IEEE 802.11w standard is in its early proposal stages.

10.   Wi-Fi and Other Technologies, Concurrency or Complementarity?

A very great interest in the use of WLAN technology, Wi-Fi and its successors has made evident along the chapter and should emerge clearly. Despite regulatory restrictions that may be present in certain countries, the combined efforts of manufacturers, standard bodies, user groups, system integrators, marketing pressure, etc. seem to have pushed the development of the WLAN market well further. Similarly, issues regarding regulation, security and roaming (IEEE

802.11a/g/h/i/e, UWB) have been addressed by specific working/task groups, so not to pose a business or technical risk for the near future. WLAN are fast becoming the "last mile" of choice for the overall Internet infrastructure. Most business offices, universities and airports, use WLANs as Internet access links. Actually, WLANs demonstrate clearly that they hold the promise of providing unprecedent mobility, flexibility and scalability than their wired counterparts.

It will certainly be necessary to clarify the positioning and strategy of mobile and fixed operators, traditional and wireless internet service provider (ISP), to figure out where the wireless sector is heading and next challenges.

From a user point of view, what it matters is always the answer to (at least) the simple questions – how fast, how far, how much – in order to make up his mind for new implementations, technologies and budget allocation.

Wi-Fi networks evolved in terms of rates, QoS and security; they can in some cases compete with the cellular mobile networks and in other cases being complementary with other technologies wireless/wired/mobile. Among other technologies which are born, we can quote, for example, WiMAX (IEEE 802.16) and WiBro (derivative of WiMAX). While Wi-Fi networks fulfil the need for data services in local areas, IEEE 80.216 aims at serving the broadband wireless access in metropolitan area networks. We find also IEEE 802.20 which proposes similar functionalities; its objective is to integrate in future versions of the cellular networks a mobile solution based on the OFDM. On another side, the distribution networks of television are seeing standards like DVB-T or DVB-H (cellular phones) emerging and which will revolutionize the distribution of images. At shorter distances, technologies like Bluetooth, UWB and ZigBee should not be forgotten.

It is clear that the future will continue to be rich in evolutions of wireless networks.

Chapter 3

Bluetooth$_{TM}$: Architecture and Functions

1. Introduction

Bluetooth technology aims at allowing wireless short-range communications between several devices. Developed originally by Ericsson, Bluetooth undergoes an evolution of its specifications maintained and developed by the Special Lobby Special Interest Group (SIG) of Bluetooth and is accessorily standardized by the IEEE under the reference IEEE 802.15.1. Today IEEE 802.15 subgroups and other forums such as the Wimedia Alliance are competing for the same field of operation.

The governing idea behind Bluetooth consisted in specifying a wide scale integrated circuit to be deployed on a very large scale on various types of equipments with a very reduced energy consumption and thus announcing very low prices (Figure 3.1).

1.1 SIG

In 1994, Ericsson Mobile Communications launched a feasibility study of low-cost low-consumption radio interface to be used between mobile telephones and their accessories. In February 1998, IBM, INTEL, Nokia and Toshiba joined the Swedish company and in May they created the SIG. It was widened by the arrival of 3Com, Agere (Lucent Technologies), Microsoft and Motorola during 2000 and does not cease increasing thus gathering actors who cover several fields of expertise such as cellular telephony, portable computers, cars

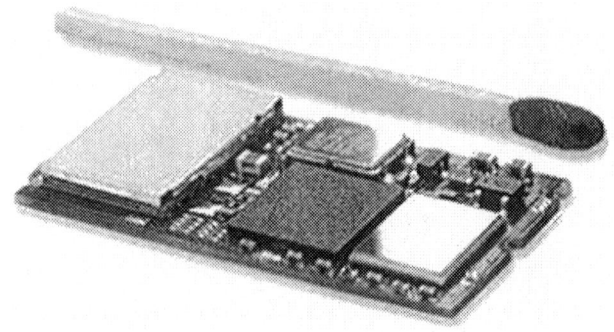

Fig. 3.1. Bluetooth chip

and digital processing. Being an opened industrial specification, all the members of Bluetooth SIG can use it free in their products and services.

Today, the SIG counts more than 2500 manufacturers. It carries out a true fight to promote this world standard despite the large number of concurrent technologies essentially headed by Wi-Fi.

1.2 Bluetooth Does it Relate to Teeth by Any Sense?

Bluetooth comes from the name of a Viking: *Harald Blätand* ("Harald the blue tooth") who succeeded the exploit of unifying within the same kingdom, Denmark and Norway, at the time where Europe was divided by religion quarrels and territories. The metaphor is beautiful: today Bluetooth carries out a true crusade of modern times to unify the whole of the manufacturers and their electronic instruments thanks to this new wireless standard. The icon is also based on the Bluetooth history because it is composed of two characters "H" and "B": abbreviation for "Harald Bluetooth".

Bluetooth products have to go through a certification process to become the group products.

1.3 Applications

Mobile telephony is one of the first markets concerned by Bluetooth. Bluetooth chips are easily added in cellular equipments. Another type of application relates to the data-processing equipment such as mice, portable printers, PC, organizers, audio helmets, Hi-Fi systems, numerical cameras, remote banking payment and car applications. We call an application a *profile*. The profile gives details about protocols parameters and settings for device to be able to discover these applications and to communicate in a uniform way. Profiles are specified in the SIG.

2. Architecture and Throughputs
2.1 Architecture

Bluetooth communication requires two preliminary things: first we have to know the devices in the neighbourhood (discovery) and second there must be a pre-established circuit. Communication is also based on a master–slave principle. A group of equipments forms a cell called piconet (see Figure 3.2). A piconet comprises a master and seven slaves at the maximum. Several piconets can overlap and form a "scatternet" (see Figure 3.3). In a piconet the communication is based on the master to harmonize the frequencies and channels. We know the neighbours through the discovery phase while in a scatternet there is a need to route data between masters and relay nodes. Scatternets in Bluetooth is not well developed. It has been improved by specific routing procedures in later standards such as ZigBee.

Two slave devices cannot talk directly to each other except during the discovery phase. Channel allocation and communication establishment are under the responsibility of the master.

Although there was a limitation in earlier versions of Bluetooth on the number of simultaneous channels in a piconet, it is removed from the current version as the cell capacity has increased significantly. The standard supports also broadcast by simply removing the destination from the messages.

The master is responsible of polling nodes and also allocating/blocking new connection bandwidth. It is responsible for setting the piconet synchronization clock and as we will see decides for the frequency hopping sequence (FHS). A slave can be part of several piconets.

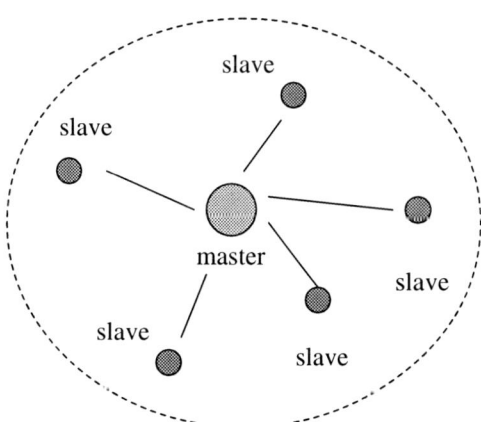

Fig. 3.2. Master/slave architecture

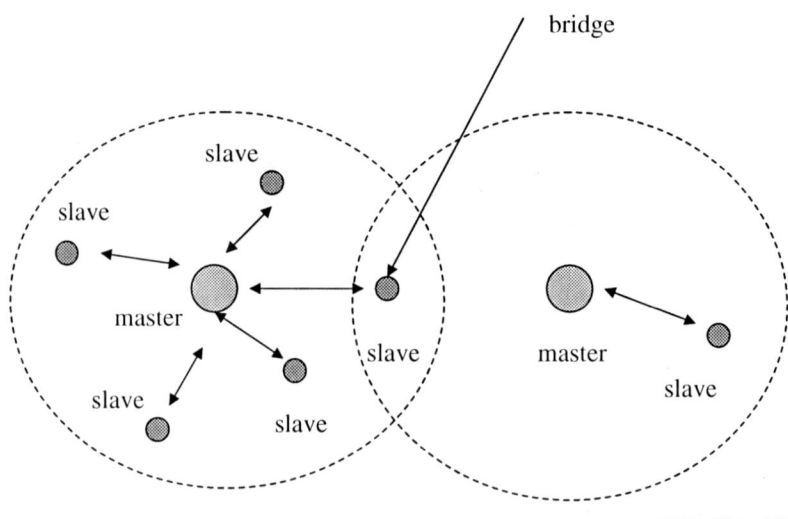

Fig. 3.3. Scatternet

One major interesting feature of Bluetooth is that it is not dependent on the IP. This courageous design decision eases the deployment of devices that do not need to worry about upper layer problems such as address allocation, default router, netmask, etc. Auto configuration is hence much easier. In Bluetooth we identify several protocols:

- Lower layer protocols: *Baseband*, LMP, L2CAP, service discovery protocol (SDP)
- Interfacing protocols: RFCOMM
- Applicative control specifications: TCS Binary, AT Commands
- Applicative protocols: PPP, TCP/IP, OBEX, WAP, vCard, VCal, WAE

As we can see in Bluetooth, applications are part of the general specifications. As mentioned before they are defined in separate documents defined by the SIG.

For a better understanding of the Bluetooth mode of operation, we will look at the protocols shown in Figure 3.4 and their goals.

2.2 Throughputs and Versions

Several versions of the specifications exist. Version 1.0 offers a maximum bandwidth less than 1 Mbps for a range of about 10 m. The range depends on the equipment class. Three classes and hence three transmission powers are defined for very low, moderate and average power devices (e.g. small equipment, telephones and computers cards). Version 2 offers a maximum bandwidth ranging

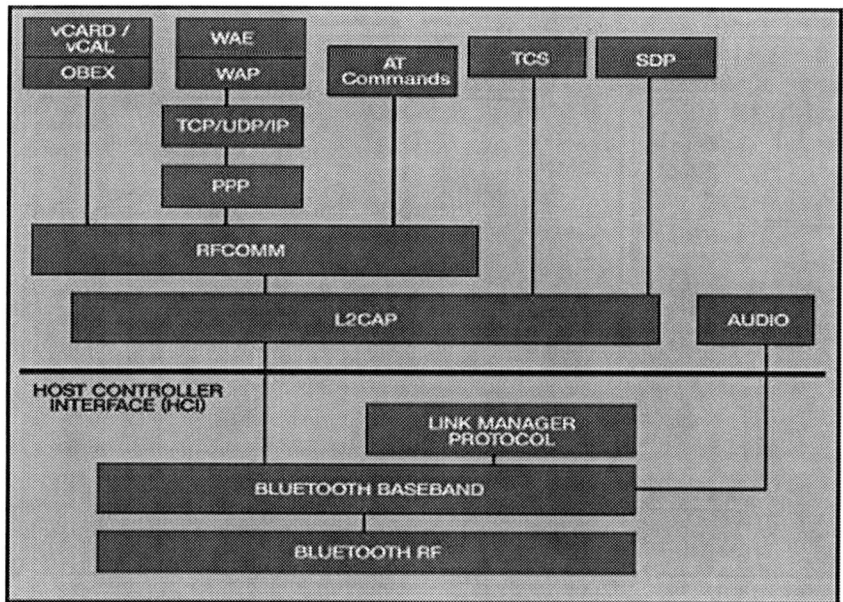

Fig. 3.4. General stack architecture

from 2 to 3 Mbps. When Bluetooth goes to higher ranges it faces other standards such as Wimedia and is hence not necessarily in a favourable concurrent position.

In this chapter, we refer to 1.1, 1.0B and 2.0 specifications.[1]

3. Physical Layer and Physical Channels

The physical layer is based, for the basic rate, on frequency hopping with GFSK modulation. The modulation produces 1M symbols for a total throughput of 712 Kbps. In this kind of modulation we generate a binary one by a positive frequency deviation, and a binary zero by a negative frequency deviation. The advantage of this technique is that it does not require complex demodulation and RF front-ends. A very simple system is hence required. An example of such a system is depicted in Figure 3.5.

The frequency hopping is used as a very primary spreading spectrum technique. It has been kept in the new version of Bluetooth. We use 79 channels in the standard but sometimes, according to each country, we may use less channels and hence have globally less bandwidth.

---

[1] Different versions of Bluetooth specifications (1.0, 1.1, 1.2, 2.0) are available at the official site of bluetooth http://www.bluetooth.org.

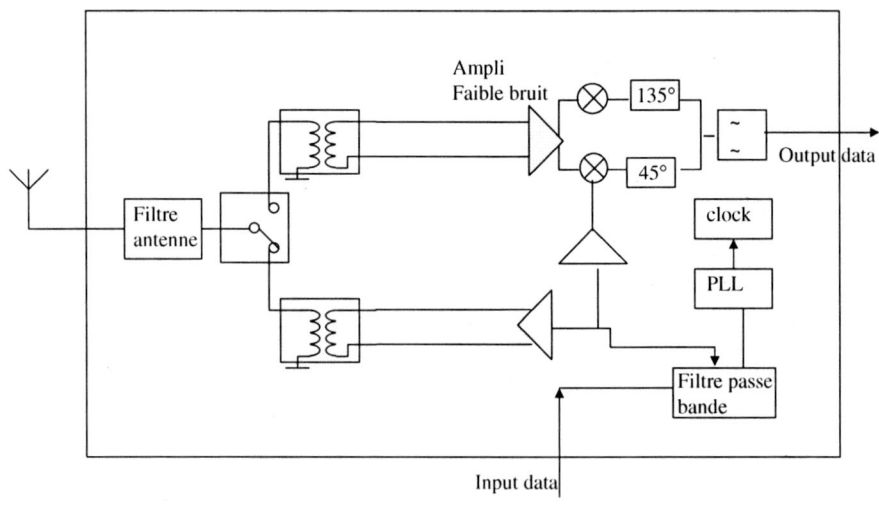

Fig. 3.5. A simple chip, GFSK modulation

Table 3.1. PSK modulation, 2 Mbps rate

Bit(2k-1)	Bit(2k)	Φ_k
0	0	$\pi/4$
0	1	$3\pi/4$
1	1	$-3\pi/4$
1	0	$-\pi/4$

In the enhanced rate version, we use a different modulation. However, the header is always sent at the same low speed and uses GFSK. This is a principle also found in the WLAN 802.11 technologies. It enables devices to adapt their RF ends to the dynamic modulation technique for each frame.

The new PSK modulation reaches a bandwidth of 2 Mbps. It is based on 90° phase shift keying ($\pi/4$-DQPSK). The 3 Mbps is based on 8-ary phase shift key (8DPSK). Table 3.1 shows relations between received bits and the detected change of phase in the 2 Mbps case.

Four types of channels are defined in the standard definitions. Two are used for normal data transmission and are called basic and adapted piconet channel. A simple device is required to use one channel only at a time but enhanced devices could do parallel multiplexing over several physical channels.

The frequency hopping requires that the change be 1600 hops/s in data transmission and 3200 for inquiry and paging. A channel is hence associated to the hopping sequence, i.e. how we are going to change the frequencies in time. The master tells the clients about their own hopping sequence. Hopping in frequencies insures that we can bypass some frequencies that we believe they encounter a specific noise or propagation problems. It is also dependent on the clock that is mandated by the master. Data packets are sent over time slots. A time slot was fixed in the earlier Bluetooth specification. Now we have different time slot lengths. Also a packet can be sent over several time slots. Like any other layer two protocol, Bluetooth uses a special sequence to delineate the beginning of the packet. This is called an *access code*.

The adapted channel is only different in the hopping sequence procedure. It requires to use at least 20 channels. It needs also that the slave uses the same frequency to answer back that was initially used by the master. It is however less restringing in the fact that we may use less than 79 channels for the sequence.

3.1 Frequency Bands and RF Channels

Bluetooth operation is in the 2.4 GHz ISM band. In most countries we have 83.5 MHz that should be dedicated to this protocol. 79 channels are hence possible in this range with a bandwidth of 1 MHz per channel. Table 3.2 gives some restrictions in different countries.

As mentioned before we have three different classes of devices:

- Class 1: it is designed for high-range devices such as Bluetooth Access points ($\sim$100 m), power characteristics: 20 dBm (100 mW)

- Class 2: for normal PCs and portable plugged devices ($\sim$10 m), power characteristics 4 dBm (2.5 mW)

- Class 3: for low power devices (range less than 1 m), power characteristics: 0 dBm (1 mW)

These power values are measured at the input to the antenna. Nothing prevents a device to vary dynamically its power. The receiver should have a sensitivity of -70 dBm to detect far equipments.

Table 3.2. Frequency bands and RF channels

Countries	Frequency range (MHz)	RF channels (MHz)	
Europe* and the United States	2400–2483.5	f=2402+k	k=0,...,78
France	2446.5–2483.5	f=2454+k	k=0,...,22
Spain	2445–2475	f=2449+k	k=0,...,22

4. Baseband Layer

Baseband is a layer that implements a very large amount of procedures linking the data transmission to the radio coding and modulation. Clock, data packet formats, master/slave roles, connection/sleep state machine management, link control and audio transmission and forward error correction (FEC) (coding also) are defined here.

4.1 Physical Characteristics

Physical Channels. The logical channel is mapped onto a physical one. It is represented by the pseudo-random sequence hopping chosen among the 79 or 23 channels RF available in the 2.4 GHz band. The Bluetooth devices which use the same sequence form a piconet. The hopping sequence is unique for each piconet except if it is adapted for a channel. The clock of the master is to be used of course. The channel is divided into intervals of time or slots. Each interval of time corresponds to an RF frequency among the hopping sequence so that two consecutive slots correspond to two frequencies. Normally the master uses a frequency downstream to the slave and the slave uses the following slot for the upstream communication. The multiplexing technique uses time division time division duplex (TDD), i.e. the master and slaves transmit alternatively (a master transmits in the even slots and the slaves in the odd slots). The intervals of time or slots are numbered and last 625 μs each one.

A transmission can be done only between the master and a slave or a slave and his master. The direct slave–slave communications cannot take place EXCEPT for discovery. The classification of the slots corresponds to the value of the Bluetooth clock of the master. This 27 bits wide clock helps to number the slots from 0 to 2^{27-1} in a cyclic way (Figure 3.6 and 3.7).

The data transmitted out by the Bluetooth units are using packets. A packet corresponds to the juxtaposition of 1–5 consecutive time slots. When a packet has a size of 1 slot, one speaks about single-slot transmission/reception, and when a packet has a size higher than 1 time slot (3 or 5 slots), one speaks about multi-slot transmission. The frequency hopping rules applied to the packets mandate that the beginning of packet will be aligned on the beginning of the slot. The frequency hopping remains fixed all the duration of the packet transmission.

Logical Links. Several types of links can be established between a master and one (or several) slave. We enumerate the most used here:

- Synchronous connection oriented (SCO)
- Extended SCO (ESCO)
- Asynchronous connection oriented (ACL)

Bluetooth: Architecture and Functions 83

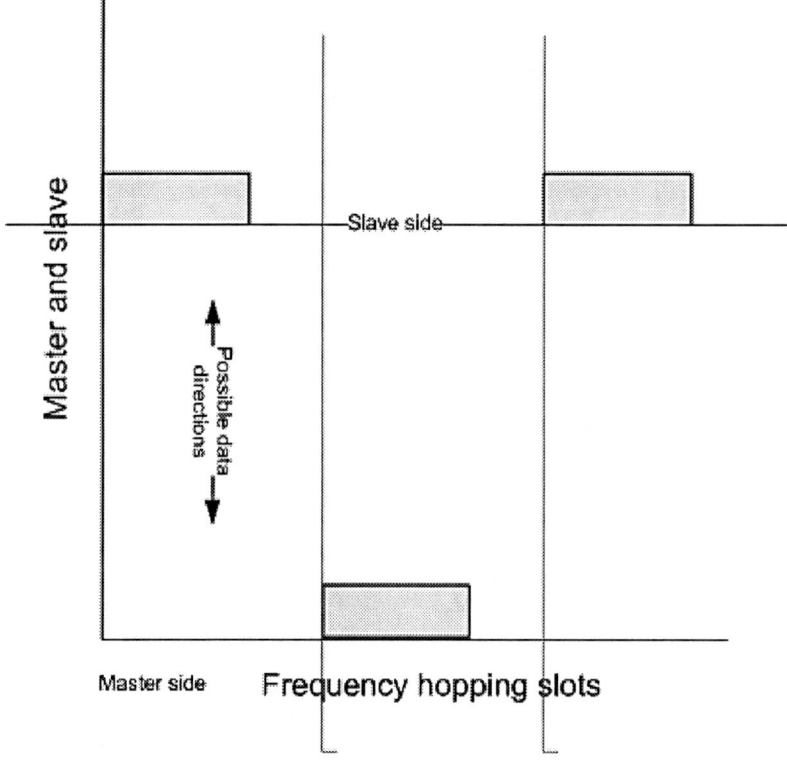

Fig. 3.6. Cyclic transmission/reception using single slot

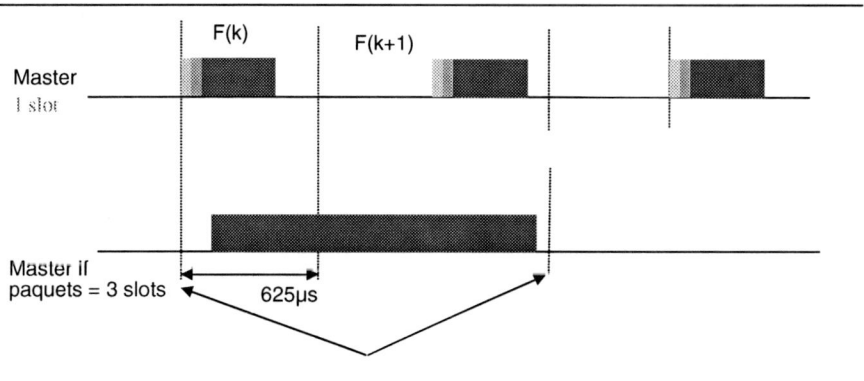

Fig. 3.7. Comparison of single and multiple slots transmission

SCO are used for:

- Voice communications in circuit mode

- Synchronous and symmetrical services

Reservation of slots at regular intervals gives 64 Kbps per slot. The SCO link is symmetric, i.e. you have same slots upstream and downstream and they are point to point.

ESCO are asymmetric links and they offer more packet types supported on their communications.

ACL links are used for:

- Data communication

- Symmetrical and asymmetrical asynchronous services

- Discovery and paging

In general, all signalling uses ACL. Also by default any slave has an ACL to communicate with the master. The maximum throughput a user can have on an asynchronous link is 723.2 Kbps for Bluetooth phase I devices. In this case, a upstream link of 57.6 Kbps can still be supported. Symmetrical flows ranging between 108 and 433 Kbps are also possible. The synchronous link supports a full-duplex connection (full-duplex) with flows that are multiple of 64 Kbps in each direction.

Link Combinations. The node manager may choose any combination of link types. The choice is left open to choose either SCO or an ESCO connections, if the application parameters allow it, an ACL connection may be simultaneously used (it is always there). There is however an important condition that states that nothing can happen before the node gets connected and this cannot happen before it establishes a primary ACL connection with the master as stated earlier. The logic behind it is clearly to enable the node to receive commands from the master at any time.

4.2 Addressing

Four types of addresses are used between Bluetooth devices: BD_ADDR, LT_ADDR, PM_ADDR and AR_ADDR:

- BD_ADDR: Bluetooth device fixed address. It is provided by the manufacturer. Each device must have a unique address. The length is 48 bits. It corresponds to an MAC address. It is however as we will see only in signalling to start a communication but never after. The funny thing about

MAC addresses is that they represent the only "unchangeable" device identification. Any other address can be changed. Of course this raises the question about the difference between the identity and the address that are here the same (while they should be different), but in the lack of any provable fixed identity an MAC address remains a good solution.

- LT_ADDR: the address of the active Bluetooth member. It is a 3-bit wide number and it elucidates the secret behind the maximum number of eight nodes in a piconet. This address is only valid as long as the slave is active in a piconet. This address along with packet type identify the required connection in a slave. Note that the slaves do not communicate directly between each other, hence this address is only to identify a slave in the downlink communication.

- PM_ADDR: it is an address reserved to the non-active members. This address is on 8 bits and is valid only if the slave is parked. The PM_ADDR is assigned to the slave by master during the parking procedure.

- AR_ADDR: This access request address (AR_ADDR) is also temporary. It is used by parked (sleeping) slaves to determine the next slot that it is allowed to use to send a new access request. It is assigned by the master to the slave and again is valid as long as the slave is parked.

4.3 Bluetooth Packets

There are several kinds of packets that are related to the different link types. According to the application, we identify globally three kinds of packets.

- *Control packets*: they are used between the server and slaves to carry information related to the radio (frequency hopping, clock) to the link (establishment, security) or to the connection (connection requests).

- *Synchronous packets SCO*: they are used by the connection oriented links to carry voice.

- *Asynchronous packets ACL*: they are used for ACL connections, so mainly best effort data traffic.

For each one of these types, several subcategories exist. The various types of packets which result from this list are shown hereafter.

- *Voice data packets DV*: the DV packet is a combined data – voice packet. It has a mixed payload divided into a voice field of 80 bits and a data field containing up to 150 bits. DV is for SCO links.

- *DM x packets (x = 1, 3 or 5)*: the DMx packets are for data only. They are hence used on ACL links. The payload has a 16-bit CRC code. The DM1 packet holds on one time slot and can take up to 18 bytes. Data field and CRC are coded with a 2/3 FEC (FEC is explained later). DM3 holds on three slots and DM5 on five slots.

- *DH y packet types (y = 1,3 or 5)*: it is similar to the DM1 packet. There is no FEC this time. Data and header have a 16-bit CRC. Again DH are on ACL links, the DH1 packet occupies a single time slot. There are several kinds of the DH family: DH3, DH5 (341 bytes).

- 2-DH1, 3-DH1 packets are similar to the DH1 packets except that the payload is modulated using $\pi/4$-DQPSK and 8DPSK. The 2-DH1 packet is 56 bytes long, while 3-DH1 has 85 bytes.

- 2-DH3 packet is again similar to the DH3 packet except that the payload is modulated with $\pi/4$-DQPSK. It can hold up to 369 bytes. 3-DH3 has a length of 554 and 5-DH5 has a length of 1023 bytes.

- HV1 packet has 10 bytes of data protected with a 1/3 FEC rate and no CRC. The payload length is fixed at 240 bits. There is no payload header present. It is kept simple for simple audio devices such as headsets. It is also used only on SCO links.

- HV2 packet is 20 bytes long and has a FEC rate of 2/3 FEC and no CRC. HV3 packet has 30 bytes.

- EV3 packet is used on ESCO only. It is between 1 and 30 bytes and a 16-bit CRC code. The bytes are not protected by FEC. It holds on a slot and does not have a header. The length is set in the negotiation.

- EV4 packets are 120 bytes long on three time slots. While EV5 packets are 180 bytes long.

- 2-EV3 and 2-EV5 packets use $\pi/4$-DQPSK and span up to 360 bytes.

- 3-EV3 uses 8DPSK and is 90 bytes long on a single time slot.

So in general when we have packets with their type ending with a 1: they use 1/3 FEC

- With a 2: FEC is 2/3

- With a 3: no FEC

- With a 5 : no FEC, CRC 16.

In all, there are five types of control packets (ID, POLL, NULL, FHS, DM1) that are used for control of the channel and the previously described ones for data transmission. There is an ID packet, which is used for the discovery procedures (Inquiry) and paging (Page). The two most important packets from the connection establishment point of view of link are described hereafter. They are: ID and FHS packets used for discovery and frequency hopping.

ID Packet It consists of the device access code (DAC) or inquiry access code (IAC) and has a fixed length of 68 bits. Access codes are explained later and consist in bit strings that identify devices in a short way.

DAC is 68 bits long. It contains the node address or an inquiry code to ask for a node with specific conditions. The packet is used during inquiry phase.

FHS Packet The FHS packet contains the MAC address of the sender (Figure 3.8). It is coded with 2/3 FEC rate. This packet is used in the answers to the page requests and Inquiry ones before the creation of the PICONET or when there is a role change. The CLK27-2 of this package contains a part of the clock of the transmitter. This field is updated at each transmission of the package. The total size is 240 bits ($[144 + 16]*3/2 = 240$).

The CLK field contains the emitter part (of course a truncated part).

Packet Formats. As it can be understood up to now, we can see that all data categories are sent in packets (even for streamed audio). Hence, the general packet format has been divided into basic rate packets and enhanced rate ones (Table 3.3). For the basic rate the structure is as follows:

Three parts that are explained hereafter.

The access code has two possible sizes, respectively 68 or 72 while the header is fixed with 54 bits, the payload contains the data and is at maximum 2745 bits. In the previous packet ID type we saw that the packet is quite small. In fact a packet can be short with only an access code. It can also

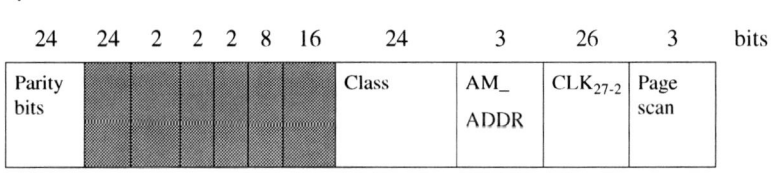

Fig. 3.8. FHS packet structure (payload part)

Table 3.3. Bluetooth general packet format

68/72 bits	54 bits	[0–2745 bits]
Access code	Header	Data

Table 3.4. Bluetooth general packet format for enhanced rate

68/72 bits	54 bits				
Access code	Header	Guard	sync	Data	Trailer

4 bits	64 bits	4 bits
Preamble	Sync word	Trailer

Fig. 3.9. Access code format

be composed of the access code and the packet header or the access code, the packet header and the payload which is the normal case.

In the enhanced rate the structure is shown in Table 3.4.

Enhanced data rate packets access code and packet header are identical in format and modulation to the basic rate packets. The structure has an additional guard time and synchronization sequence. The guard time is a period starting at the end of the last GFSK symbol of the header and ending at the start of the reference symbol of the synchronization sequence and shall be between 4.75 and 5.25 µs. The synchronization consists of a symbol that has a length of 11 µs. It is followed by ten DPSK symbols. The trailer is also a symbol equal to zero (00 followed by 00 for the $\pi/4$-DQPSK and 000 followed by 000 for the 8DPSK.

The first 72 bits are used to synchronize different piconet equipment. The role of the header is to:

– Number the packets and to detect a loss

– Give the local piconet number of the node

– To indicate the packet type

– To require an acknowledgement

– To check the content via a CRC

The rest of the packet contains the data itself. When the packet is segmented on several slots, a specific indication is put in the type code.

Access Code. We have seen that data packets, ID and other control packets start with an access code (Figure 3.9). It is generally used to identify the piconet and to synchronize the reception. The packet is not accepted until the access code corresponds to the master identity. This is a simple way to identify the

Bluetooth: Architecture and Functions

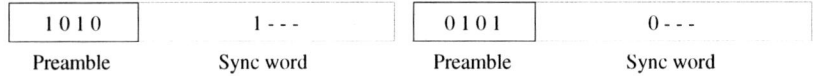

Fig. 3.10. Different values of preamble for various Sync word values

presence of several masters in a location. It can help also to try to establish a scaternet.

All the piconet packets use the same access code.

The access code in basic rate contains 72 bits if it is followed by a header and 68 bits if it is transmitted alone. It contains three parts: a header flag, a synchronization flag and a trailor flag. The three types previously mentioned are:

- Channel access code (CAC)
- Device access code (DAC)
- Inquiry access code (IAC)

The type used depends on the mode of operation of the Bluetooth apparatus. The CAC identifies a piconet. It is inserted in all the packages transmitted in a piconet. The DAC was already presented and is used for the creation of a new connection (in the Page and Scan requests) or for the transmission of special indication. IAC is used for the inquiry of other devices.

IAC also exists in several versions. A general version (general inquiry access code [GIAC]) which indicates to all the devices which receive it that they must answer, and for nodes with the equivalent class (device inquiry access code [DIAC]). It allows to seek only one class of devices. In this case, all the devices that receive the DIAC corresponding to the required class answer the message.

Preamble The preamble is a pattern of four symbols used to facilitate DC compensation. It is the same principle in most protocols to identify the beginning of a packet. They use a different pattern as shown in Figure 3.10 for each subsequent preamble according to the Sync word.

Sync Word The sync word is 64 bit long. It is derived from a MAC address (bits 0–23). In the case of a CAC we use the master address. While in GIAC and DIAC we use specific addresses. For the DAC the slave address is used. The role of the Sync word is to provide a precise timing synchronization with the sender. The Sync words are created to have a large Hamming distance between them.

Trailer It is used after the Sync word only if there is a header in the packet. The trailer consists of four symbols. It is logically there because we do not know the

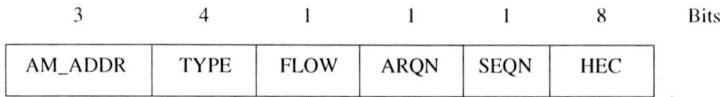

Fig. 3.11. Header format

Table 3.5. Some Codes used in the «Type» header field

Code type	Type of packets
0000	NULL
0001	POLL
0010	FHS
0011	DM1
0100	DH1
0101	HV1
0110	HV2
0111	HV3

length of the packet in the radio part and we need something to indicate it. The higher layers know the length from the header but the radio is not sufficiently smart to extract this information which is a data field in the packet.

Packet Header. The header contains six fields of information (Figure 3.11). Its size is 18 bits, a CRC and 1/3 rate FEC. This creates a packet field of 54 bits. The header is as follows:

AM_ADDR (*Active Member Address*): Address of an active Bluetooth device in the piconet.

This field is 3 bits long. It contains the slave address. This address is used in all communications with the master. It is used as long as the device is active and it is given away when the device goes to the parked mode.

The 000 address is a broadcast one.

Type It defines 16 possible packet types. It will help also to decode the information in different situations like SCO, ACL or ESCO connections. When a packet is on several slots, this gives the opportunity to others to stop listening in these periods.

Flow The FLOW bit is used for flow control of ACL packets. It helps a receiver to tell the master that its buffer is full (Table 3.5).

Flow = 0 means that the buffer is full. It is the STOP bit.

There is another FLOW field in the upper layers (L2CAP). It concerns the logical links and is a per logical link indication so it could stop the transmission over a link while other logical ones still communicate normally.

ARQN It informs the source of the positive or negative reception of a good CRC message. It is sent in the following message to acknowledge a received one. When sending one we mean the packet was received and zero means that it had some problems.

SEQN It is a bit for simply numbering packets modulo two. It is hence possible to make the distinction between even and odd messages and thus to detect repeated/lost messages.

Header Error Check. This 8-bits long field detects errors in the header and corrects them.

As stated the header consists of 18 bits with 1/3 FEC encoding giving a total result of 54 bits.

Data Field. We distinguish mainly two applications:

- Voice (synchronous transmission)

- Data (asynchronous transmission)

With the Bluetooth new version we have two kinds of synchronous (SCO and ESCO). Hence, we have a total of 15 packet types per class of link.

ACL links are used for data and SCO for voice. Only DV packet is hybrid.

Voice Information The fields in voice packets have a fixed length: 240 bits, for example, in HV and 80 bits for DV ones.

Data Field It is divided in a similar way as in normal ACL packets (Figure 3.12 and 3.13).

The payload header for asynchronous data starts with two bits called logical link identifier (LLID), a bit controls the flow as previously explained, the length indicator is 5 bits long (may be 10 bits for longer packets). The remaining bits are for further study.

2	1	5
LLID	FLOW	LENGTH

Fig. 3.12. Header of an ACL data packet when transmitted on a single slot

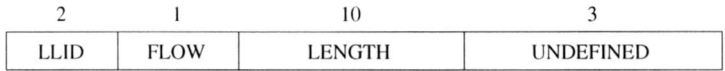

Fig. 3.13. Header of an ACL data packet when transmitted on several slots

The LLID contains a value that gives an indication on the nature of the message. It can indicate the presence of messages from the LMP layer or from L2CAP one.

4.4   Error Control

When present error control field (CRC) is 16 bit. The payload header body fields are used to generate the CRC code.

Error Control Codes.   Some packets are protected from transmission errors. Before coding, the header and data fields are scrambled to avoid the too frequent appearance of certain continuations of bits. This jamming is carried out before the calculation of the FEC. Data is scrambled by carrying out an XOR with a jamming word.

As shown before, Bluetooth has three kinds of FEC.

Header Error Check.   Header error check (HEC) is an 8-bit CRC at the end of the header. To check if the packet is correctly transmitted, it is calculated at the receiver. HEC system is initialized in different ways according to the kinds of packet. The HEC does not make it possible to correct errors, only to detect them.

CRC.   It is generated in the same manner as the HEC. It has a result word of 16 bits concatenated at the end of the data body. As for the HEC, the checking procedure can be done by calculating the CRC pattern of the payload. If it is 0, the packet was correctly transmitted, if not the packet is eliminated because it contains one or more errors.

1/3 FEC Coding.   Each bit is simply repeated three times (Figure 3.14).
Header and some voice packets use it (HV1).

Decoding is done by calculating the average of the values of the received bits, three bits by three bits, if it is higher than 1., the received bit is 1, if not it is 0.

2/3 FEC Coding.   It is a Hamming code with (15,10). The Hamming distance is 4. It is used for many packets such as DM, DV, FHS and HV2.

Bluetooth: Architecture and Functions 93

Automatic Repeat Request (ARQ). The data field in packets like DM, DH and FD are retransmitted until a positive acknowledgement is received or if timeout is reached.

An ACK (ARQN = 1) or a NAK (ARQN = 0) corresponds to the positive or negative answer related to the transmission of the previous packet. If the timeout is reached to state machine simply discards the packet and proceeds with the next one.

Different Kinds of FEC for Packet Types (Summary). Table 3.6–3.8 give these details.

Audio encoding

A last sentence before going to the upper layers has to be said on audio. Bluetooth uses old audio codecs with 64 Kbps log pulse-code modulation (PCM) format where (A-law or µ-law) may be used. It has also continuous variable slope delta modulation (CVSD) encoder offering 64 Kbps. It is to be said that we can also use upper layers with VoIP methods. Such a solution is not good in this protocol stack since we will inherit of the huge number of useless payloads probably unnecessary between a portable device and its earsets.

5. Link Manager Protocol

A link manager sets up, authenticates and configures the link. It discovers other devices and communicates with them by the Link Manager Protocol (LMP). By the first LMP message we can, for example, know if the peer can support special features such as receiving segmented packets over several slots, able to become a master and so on (Figure 3.15).

To carry out its role of supplier of service, the management protocol employs the services of the link controller (LC) through a whole set of messages. This

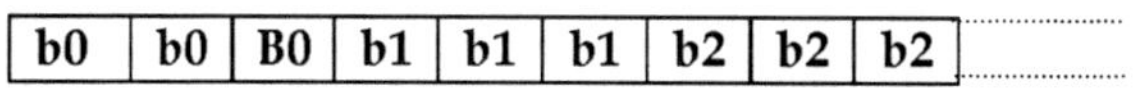

| b0 | b0 | B0 | b1 | b1 | b1 | b2 | b2 | b2 |

Fig. 3.14. 1/3 FEC coding

Table 3.6. Control packets

Type	Data length (octets)	FEC	CRC	Max symmetric bandwidth	Max asymmetric bandwidth
ID	–	–	–	–	–
NULL		–	–	–	–
POLL	–	–	–	–	–
FHS	18	2/3	Yes	–	–

Table 3.7. ACL packets in basic rate

Type	Header size	Data size (bytes)	FEC	CRC	Max symmetric bandwidth (Kbps)	Max asymmetric bandwidth	
DM1	1	17	2/3	Yes	108.8	108.8	108.8
DH1	1	27	No	Yes	172.8	172.8	172.8
DM3	2	121	2/3	Yes	258.1	387.2	54.4
DH3	2	183	No	Yes	390.4	585.6	86.4
DM5	2	224	2/3	Yes	266.7	477.8	36.3
DH5	2	339	No	Yes	433.9	723.2	57.6
AUX1	1	29	No	No	185.6	185.6	185.6

Table 3.8. SCO packets

Type	Header size	Data size (bytes)	FEC	CRC	Max symmetric bandwidth (Kbps)
HV1	–	10	1/3	No	64
HV2	–	20	2/3	No	64
HV3	–	30	Non	No	64
DV	ID	10+(0–9)D	2/3 D	Yes	64+57.6 D

D: Hamming distance

protocol is composed primarily of a certain number of commands sent from one device to the other. LMP messages are sent with the ACL links explained previously. They are distinguished from data ACL link and their name ends with ACL-C. The difference between ACL-C and ACL-U (carrying plane data) is detected by the LLID field of the payload header.

The LMP message format starts with 7 or a 15-bit operation (opcode) that is used to identify the required operation (Figure 3.16). If we have an opcode with the values 124–127 it means that an additional byte of opcode is located in the second byte of the payload.

The transaction ID helps to identify messages (requests and responses) corresponding to one session.

As shown in Figure 3.15 the Paging procedure triggers the device to get into connected state and triggers all possible LMP messages. This is done through LMP message. There are several states:

– Standby

– Page

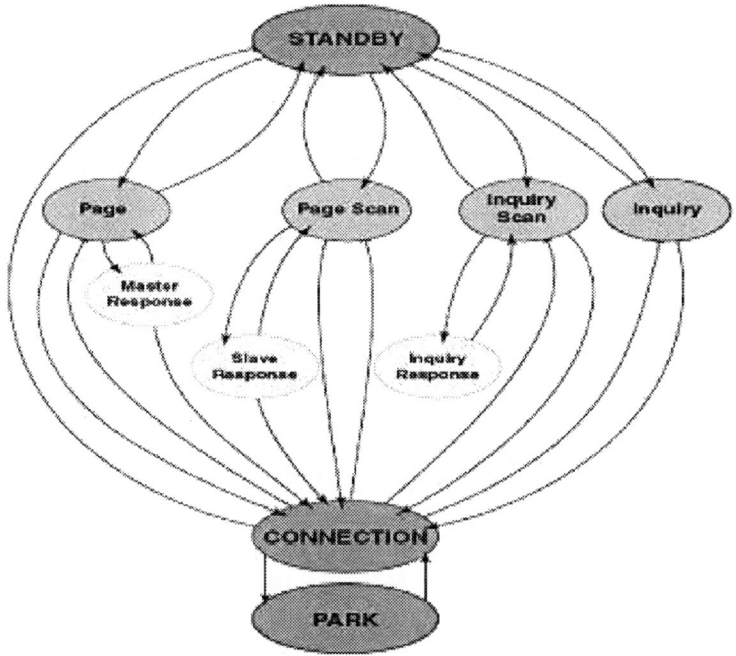

Fig. 3.15. Bluetooth state machine

Transaction ID	OPCODE	PAYLOAD

Fig. 3.16. LMP message format

- Inquiry
- Connection
- Park

Note that the Inquiry is not a state managed by the LMP, i.e. it is not due to LMP messages that we make an inquiry but a more higher layer protocol (host control interface [HCI]).

Some LMP Procedures. The LMP procedures include the connection phase, the detach phase, power control and adaptive frequency hopping (AFH). AFH improves the performance of the physical link in the presence of radio interference. It can also reduce its own interference with other external devices and it is only used during the connection state. We find also among important procedures the *power control procedure*. If the received signal changes, the Bluetooth device may request its neighbour to increase or decrease the other device TX power. These requests are made anytime after the paging procedure.

We enumerate hereafter some of the important LMP procedures to give an idea on the LMP role in general.

5.1 LMP_Sniff_Req, LMP_Unsniff_Req

The *sniff* mode enables nodes to get into a slower cycle and to save energy, the time slots when a slave is listening are hence reduced. The master transmits to a slave in specified time slots. Either the Master or slave may request to enter in the sniff mode. They both negotiate an interval of time T for discovery and an offset. This message is hence sent to start the sniff procedure. The receiver may then accept or reject it. When a link is in the sniff mode, only the master can begin a transmission in the slots. The start parameter determines the number of slots the slave must listen, starting from the sniff anchor point. The stop parameter determines the number of slots that the slave must listen additionally.

5.2 LMP_Host_Connection_Req, LMP_Setup_Complete

The basic connection establishment procedure is the connection request. It happens after paging a node. The destination node may accept or reject the connection by the equivalent procedures: LMP_accepted or LMP_not_accepted (Figure 3.17).

When no connection is needed the node goes automatically to the standby state. It listens periodically to messages (may be set to 1.28 s). The connection procedure can be initiated by any node that will become hence the master.

A connection is established by sending a paging message on the page scan channel. As it was explained previously page scanned periodically by the node to know if somebody is requesting a connection. This is done if the address of the unit to be connected (slave) is known or then an inquiry message followed by a Page if the address is not known. The following exchanges take place:

- Being in the Inquiry state, M sends a signal to ask E that it needs to initialize a connection. E is then in the state inquiry scan.

- E passes then in the sub-state Inquiry response then answers the master. The response of E comprises its address (48 bits MAC) as well as information on its clock–once that E sent its answer, it passes in the Page scan state. It is put on standby waiting for a message comprising its own address on one of the existing channels. When M receives the response from E, it passes into the Page state (M stores the received information). This information makes it possible for M to note the presence of E. When

Bluetooth: Architecture and Functions

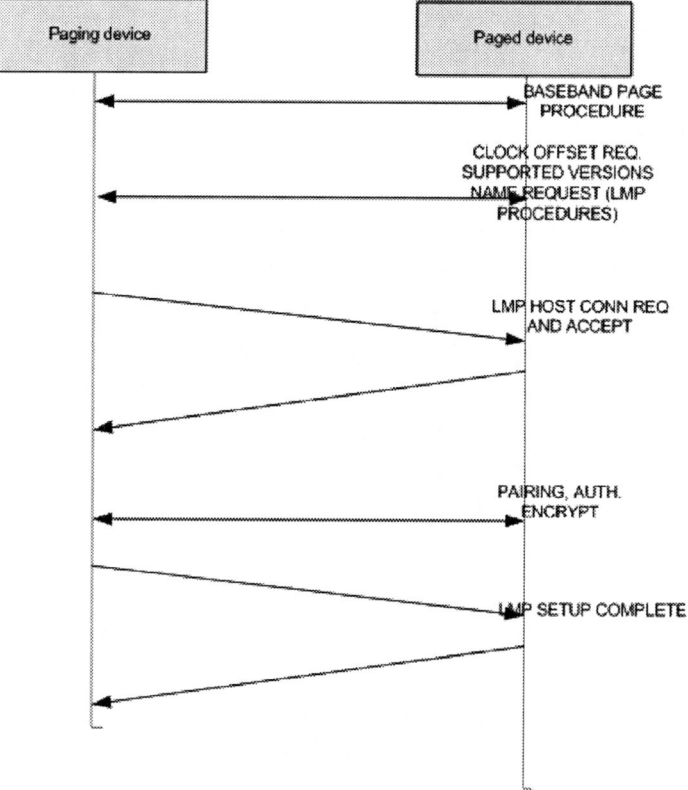

Fig. 3.17. The basic connection procedure in LMP

M wishes to continue the connection process, it returns a response message placing the E address in it. This message is returned several times on all the channels.

- When E sees an answer which is targetted to itself, it is moved to the sub-state Slave response and then returns another response message to M by joining its access code to it.

- On its side M, once this access code recovered, is placed then in a master response state and returns a packet with the FHS to E which enables him to be synchronized with M.

- Once this last message sent, M passes in the connected state. In the same way, when E receives this message it also passes in the connected state. In this case, the connected state is not a sub-state. To check that connection occurred well, the master sends a POLL packet and awaits in return for any type of packets. If a connection occurred indeed, the slave is synchronized with its master.

The connection time (time needed for the whole procedure) is typically 0.64 s (this is only valid if the slave address is already known). Since less than 1 s is required to connect a node, it is better to go into the standby mode and save batteries.

5.3 LMP_AU_RAND, LMP_SRES

Authentication is based on a challenge response mechanism where the user enters a challenge friendly called the PIN. This procedure is called pairing because it establishes a session key after a successful authentication between two devices. This is why the node starting the communication sends a command called ≪LMP_au_rand≫ that contains a random number (the challenge) to the recipient. A specific algorithm is used to combine the address, the PIN, the random number and to send back the response in the LMP_SRES that looks strangely like GSM algorithms. A detailed section on Bluetooth security is found later in the book. This is only to explain that many important procedures happen in the LMP layer.

Once the node is connected, it can be in different states as shown in Figure 3.18.

- *Active sub-state*: is a state where the node is not parked. It is normally connected or it is listening to the traffic.

- *Hold mode*: the physical link is only active during SCO and ESCO slots. All asynchronous links remain inactive. The node is operated once for each invocation and return to the previous mode. The node can sleep when not programmed for the SCO slots because it is not expected to receive anything else.

- *Park mode*: the device cannot communicate with master. It waits on a channel called public service broadcasting (PSB) to listen to the master. The slave communicates at certain periods only. It means again that out of these PSB period the slave is sleeping. A slave in this state is inactive, it does not receive and send any more messages. Its only activity consists in awaking from time to time to synchronize with the master thanks to "beacons" that are sent regularly. This passive state allows to release one of the seven places available in a piconet. More than seven Bluetooth peripherals can thus cohabit within the same piconet.

- *Sniff mode*: a slave can alternate N slots of sleep states (energy saving), and K slots of active states. Normally the sniff mode affects ACL links. They should not be used in the inactive slots. SCO and ESCO are not

Bluetooth: Architecture and Functions

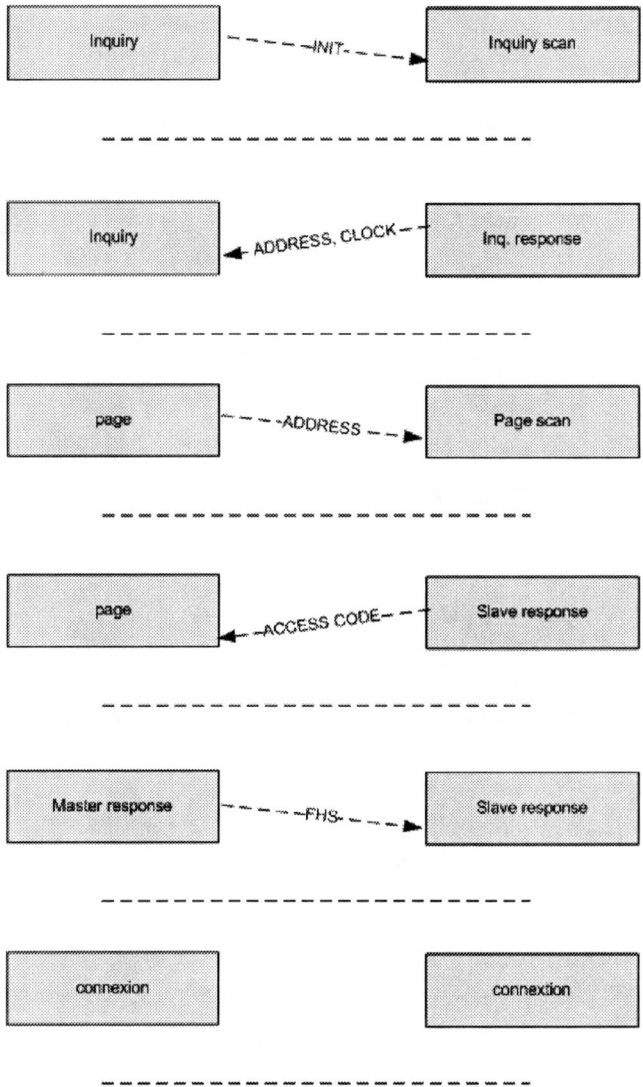

Fig. 3.18. Connection mechanism in LMP

affected since they are scheduled at given intervals while the sniff mode is still operational.

6. Logical Link Control and Adaptation Protocol

Logical Link Control and Adaptation Protocol (L2CAP) is on the same level as LMP but it still uses its premises to establish a link (Figure 3.19). It is

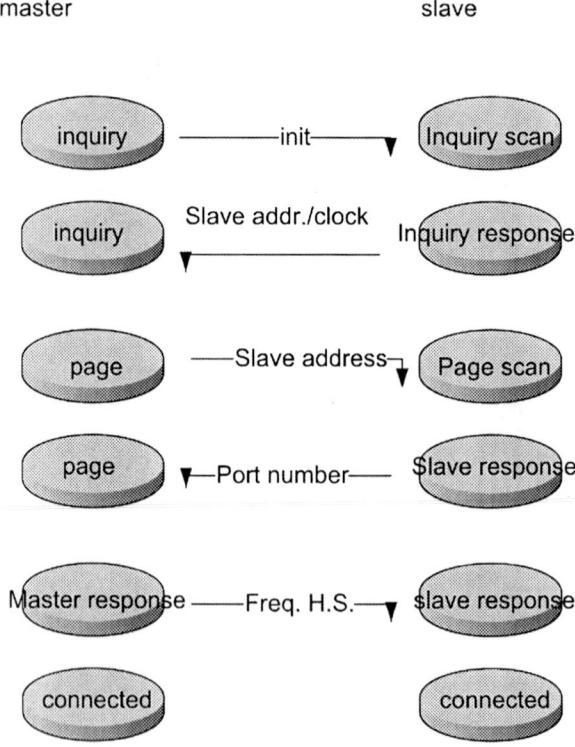

Fig. 3.19. L2CAP and LMP

not equivalent to the IP world since it is a connection establishment protocol. L2CAP is seen by applications running on devices such as telephones and PDAs. The port number for example is the same as in TCP/IP, UDP/IP protocols where it indicates the port number of the application. It enables application protocols to send PDUs up to 64 KB and implements segmentation and reassembly. A *channel identifier* (CID) helps to identify the connection in a similar way as in WiMAX protocol.

There are three types of L2CAP links:

- Bidirectional channel links for signalling.

- Data connection for bidirectional traffic.

- Unidirectionnel multicast connections. This means that L2CAP manages multicast communications for Bluetooth.

As it can be seen in 0, L2CAP uses ACL links from the baseband and does not rely on LMP (Figure 3.20). Figure 3.21 shows that we can establish logical

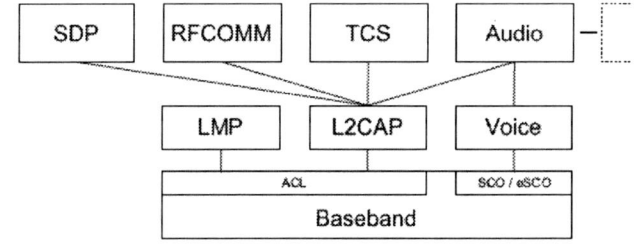

Fig. 3.20. Interaction between L2CAP and other layers

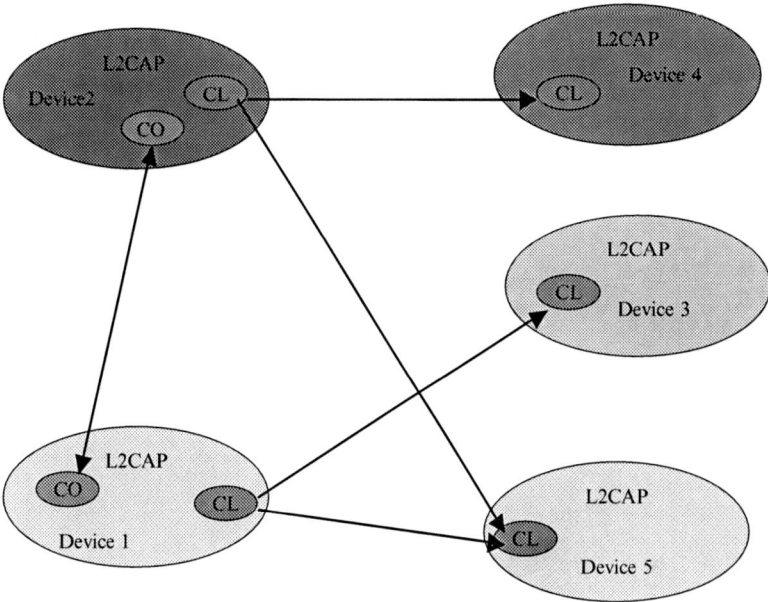

Fig. 3.21. L2CAP parallel connections

links between several Bluetooth nodes. Some may be synchronous and others asynchronous.

6.1 L2CAP Connection Establishment Procedures

L2CAP works in the original normal mode used in version 1. It is used also with retransmission and flow control. In retransmission mode the L2CAP layer implements a timer that checks for all PDUs. If not, the sender adopts a go-back-N repeat mechanism. The flow control mode is used to check when packets are not received. They are simply marked as lost.

The data transfer is not possible if the two ends of the L2CAP channel are in the OPEN state. Thus, to initialize a connection with a device, it is initially

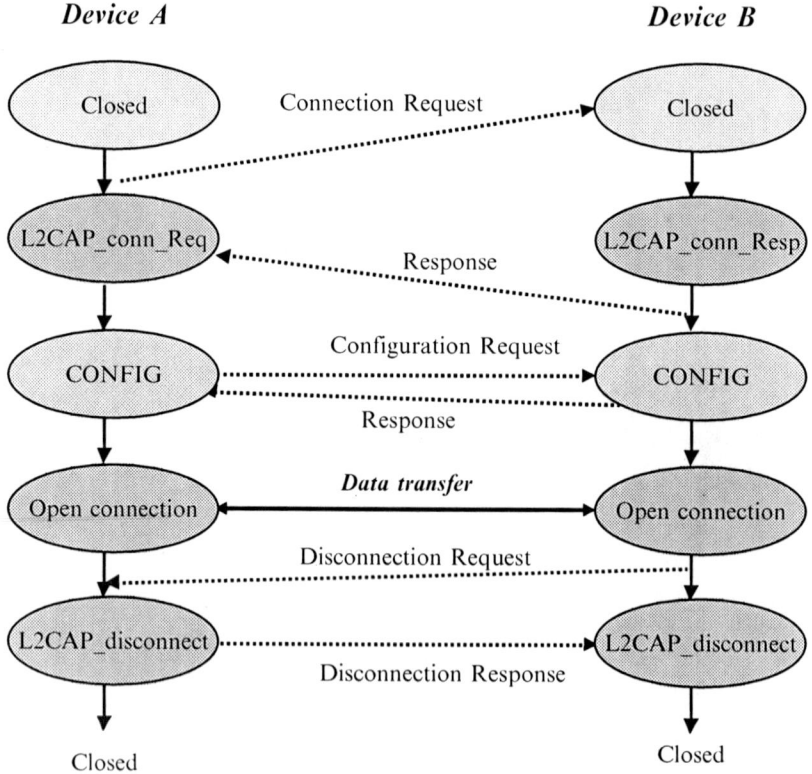

Fig. 3.22. State machine for a simple connection in L2CAP

Table 3.9. Some supported L2CAP connectionless protocols

Value in PSM	Description
1	SDP
3	RFCOMM
$0 \times F$	BNEP
0×15	UPnP

necessary to configure and connect the end point. A connection request is sent by the local device (Figure 3.22). It enters then a wait state for the response message (L2CAP_Connect_Resp). Once it receives a positive answer, it now acts to negotiate the parameters and the options of transmission (maximum size of the segments, timers, QoS, etc). It is in stage of configuration. Once all these parameters negotiated, the two entities are then in the connected state and the transfer of data can take place. The disconnection is carried out at the request of any of the two entities.

6.2 Some L2CAP Functions

Multiplexing. L2CAP has a header type different from the previous protocols.

16	16	
Length	CID	PAYLOAD

Connection-oriented L2CAP header

16	16	> 16	
Length	0 × 0002	PSM	PAYLOAD

Connectionless header in L2CAP

The difference comes from the protocol/service multiplexor used for different kinds of application protocols. Table 3.9 gives some examples of supported protocols.

Control Frame Format L2CAP supports multiplexing and is still able to distinguish between protocols SDP, RFCOMM and TCS, for example.

The commands are sent on special ACL links as explained and are called C-frames (Figure 3.23). Mode sans connexion:

The C-frame are acknowledged and checked with retransmission timers.

Quality of Service. L2CAP controls the resources to provide QoS. The process of connection establishment allows the exchange of information for this QoS established between two Bluetooth devices. Indeed, it is on the level of the configuration request and the configuration response that an option field is reserved for this use. Practically, the QoS is mapped onto ESCO or SCO links to provide the required bandwidth. In the L2CAP QoS signalling we can find leaky bucket mechanisms to implement models such as constant or variable bit rate.

Segmentation and Reassembly. As it can be seen in the baseband layer, packets are limited in size (341 octets for DH5) this is not a good solution for some applications that need larger MTUs. This is way segmentation and reassembly are supported in L2CAP.

Length	ID channel	Command field		
	(0x001)	Code	id	length
		data		

Fig. 3.23. C-frame format in L2CAP

Service Record
Service Attribute 1
Service Attribute 2
Service Attribute 3
...
Service Attribute N

Fig. 3.24. The data structure for a given service

Several packets from the baseband layer can hence be reassembled in a single L2CAP MTU or vice versa.

7. RFCOMM Protocol

The RFCOMM protocol emulates serial and USB ports over the L2CAP protocol. It uses an ETSI standard (TS 07.10) for this purpose. It supports all applications using the serial port communication. RFCOMM provides a reliable data transfer, simultaneous connections and flow control.

RFCOMM supports up to 60 open ports. An identification *data link connection identifier* (DLCI) identifies the link between a client and the application server.

The AT-Command application is used to communicate between devices supporting the modem application such as telephones, mobiles and FAX equipment. AT-Layer is directly implemented on the RFCOMM.

8. Service Discovery Protocol

This protocol is vital to Bluetooth. It is the first and only protocol that works for discovery in wireless environment. Today there are new protocols on top of TCP/IP such as JINI and UPnP, but they are really very heavy and complicated while SDP remains simple. The protocol is based on the client server model. Each node is a client and a server at the same time. Data is described in a specific attribute-based structure (Figure 3.24). SDP supports:

- Search for services by service class

- Search for services by service attributes

- Service browsing, which means to extensively look at everything supported on the device

Seven messages are defined in the protocol stack to discover and restore the information from the client to the server. Note that SDP happens after authentication since it is based on the LMP procedures.

Discovery of nodes happens before authentication while discovery of services with SDP comes later.

9. Profiles

A profile defines a set of protocol components (SDP, RFCOMM) necessary for the set up of correct and communicating applications. Today, tens of profiles are defined by the Bluetooth SIG. In general, we should look at the profile as the requested tasks that an application needs to implement to be conformant to the specifications. It defines hence several parameters in different layers of the Bluetooth stack.

We give hereafter some of the known profiles:

- *Generic access profile* (GAP), assures the good behaviour of the link layers. It describes how the devices behave from the standby to the connection state and guarantee that links and channels can be established between nodes. Discovery, connection establishment and security are described with the adequate parameters.

- *Service discovery application profile* (SDAP), The service discovery profile defines the protocols and procedures used by service discovery applications to find services in other equipment supporting SDP. This profile is not fixed because it is initiated by a user willing to find something that changes according to the scenario. We may wish to discover devices, ports or applications.

- *Cordless telephony profile* (CTP), enables cellular phones to communicate as cordless phones between a cellular and a PC. It can be seen that such a profile can make a huge difference for ADSL providers that wish to continue their communication on the cellular phone.

- *Intercom profile* (IP), offers a talkie-walkie service over devices.

- *Serial Port Profile* (SPP).

- *Headset* (HS) is one of the most used profiles. Note also that it is never implemented in the fixed PC, while it could enable the communication between the PC and the phone, and would transform the PC as a voice over IP gateway.

- *Dial-up networking profile* (DNP), to emulate a modem over a cellular phone.

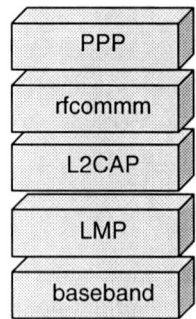

Fig. 3.25. Object transfer through OBEX protocol

– Fax FP (*Fax Profile*).

– *Local area network profile* (LAP), implements a bridging facility for Bluetooth devices to interconnect to a LAN. While very attractive, it will be shown to be very inefficient. It is also completely concurrent to Wi-Fi and is hence unused. The figure shows why it is unpractical. We will give in the implementation chapter some more details about this usage.

– *Generic object exchange profile* (GOEP), very useful profile for exchanging objects between two phones (Figure 3.25). Objects are identified by their extensions. Not all phones interpret these objects correctly. Objects could be visit cards, photos, videos, email, etc.

– *File transfer profile* (FTP), that is not related to FTP over IP, but still has the same functionalities.

– *Synchronization profile* (SP), very useful also to synchronize agendas, mailers, addresses, etc.

10. Host Control Interface

HCI provides an interface with methods for uniform access to baseband functions and independent from the upper interface. Today we have USB, PCMCIA, PCI, RS232 and WAP interfaces already defined.

Figure 3.26 shows how the interconnection is made with this protocol configuration. HCI is a mandatory implementation in Bluetooth devices acting as a master.

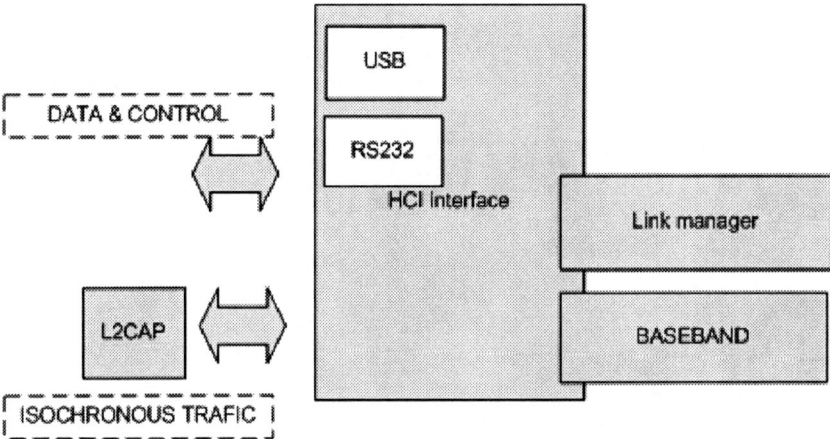

Fig. 3.26. Functional HCI entities

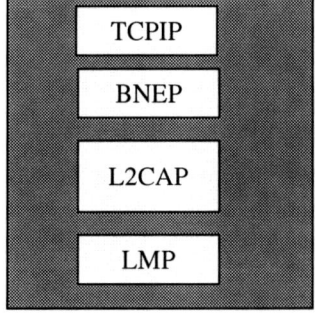

Fig. 3.27. BNEP protocol encapsulation

11. Bluetooth Network Encapsulation Protocol

A new protocol called Bluetooth Network Encapsulation Protocol (BNEP) has been added to the specifications. It is important to note that, conscious of the future of Bluetooth in the IP world, the consortium considers the improvement of the broadcast performances in this environment. For these reasons, a protocol for direct encapsulation was proposed under the name of BNEP. As shown in Figure 3.27, it allows a direct integration of IP or Ethernet in L2CAP and LMP. It supports broadcast for a better Address Resolution Protocol (ARP) that was originally implemented in a point-to-point manner in the LAN profile.

Fig. 3.28. Typical Bluetooth equipments

12. Conclusion

Bluetooth technology has certain assets: low power consumption, high level of integration, profiles, native management of the voice and it includes mechanisms for simple QoS and error control (Figure 3.28). The new version has improved data rates and makes it more attractable for high-rate transfers related to video.

Chapter 4

IEEE 802.15.4 and ZigBee™

1. General Architecture

When we talk about designing WPANs, we have three general trends that are contradictory: low data rate, medium data rate and high data rate. It is not always a desire in short-range situations to have low or high throughput because different applications require different conditions of operation. So Bluetooth can be considered as a medium data rate, while IEEE 802.15.4 is really the low data rate candidate. Unfortunately, there is no high data rate available from the IEEE but from other consortia such as WiMedia (a solution based on ultra-wideband [UWB] physical layer). IEEE 802.15.4 is designed to supply radio and MAC protocols allowing the designer to focus on the applications and customers' needs.

ZigBee is the architecture developed on top of the IEEE 802.15.4 reference stack model and takes full advantage of its powerful physical radio layer. IEEE 802.15.4 and ZigBee Alliance continue to work closely to ensure an integrated and complete solution for the market especially for sensor networking-based applications. ZigBee provides services such as security, discovery, profiling and so on for the two layers specified by the IEEE group. The target applications are summarized in Figure 4.1.

As shown in Figure 4.2 we have different topologies that can range from a centralized star or a cluster-tree-based architecture to a complete mesh network. In the last case there is a need to have an additional routing protocol. A possible architecture for mesh network is shown in Figure 4.3. Mesh networking enables

Fig. 4.1. Potential applications

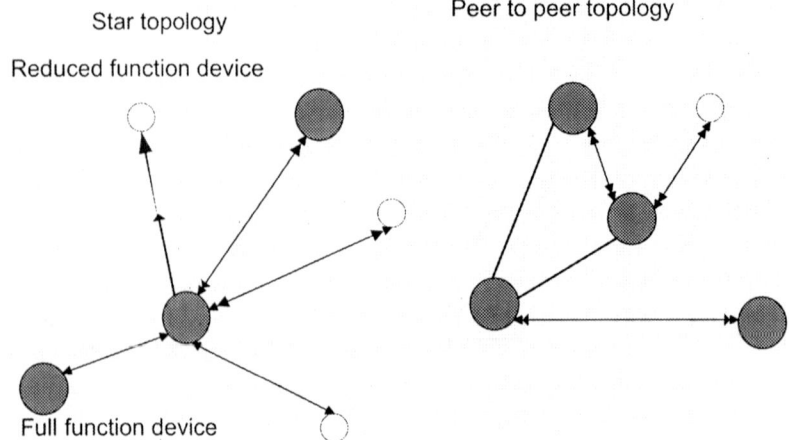

Fig. 4.2. Basic possible topologies and node categories in IEEE 802.15.4

to increase range, reliability (self-healing) and formation of ad hoc networks where redundant paths are provided.

IEEE 802.15.4 is hence a low-rate wireless personal area network (LR-WPAN) solution. It is designed to be simple for low-power devices and lightweight wireless networks. These devices rely on long-life battery normally measurable in years, but they do not claim any high throughput and should not be used in this field. As it will be shown in chapter 7, some devices such

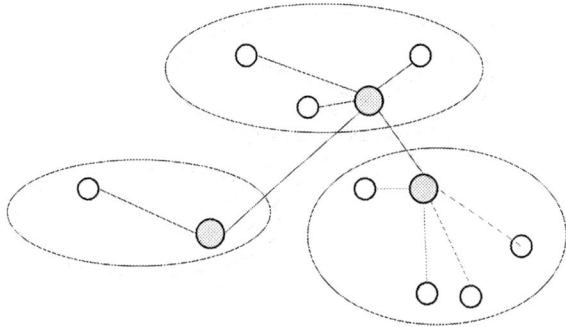

Fig. 4.3. Mesh topology with a routing algorithm

as sensors have adopted this standard and benefit from good performance in different fields such as military, environment, industry, commercial and home applications.

The data rates and features available in the initial version are:

– Data rates from 20 to 250 Kbps

– Different topologies such as conventional star and mesh operation

– Addressing based on short 16 bits or normal MAC (64 bits) addresses

– Support of simple access and slotted allocation with guarantees

– Support of acknowledged data transfer, and an optional beacon structure

– Energy detection (ED)

– Link quality indication (LQI)

– Multilevel security.

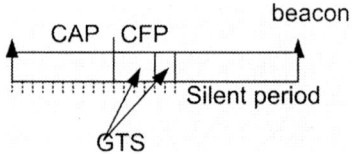

Fig. 4.4. Different access and silence periods in IEEE 802.15.4

There are 16 defined channels in the ISM band, we have 10 channels in 915 MHz band, and 1 channel in the 868 MHz band. Usage of these channels depends on each country regulations.

The standard provides specifications that depend on the device capability. For that purpose there are two kinds of capacities: a full-function device (FFD) and a reduced-function device (RFD). Normally the FFD can be a master (coordinator) and a PAN coordinator while the reduced-function one is normally a slave or simply a regular node if there is no master–slave communication.

ED is a ranging mechanism that is triggered by the upper layers to get information from the physical layer about the quality of the link.

The LQI is an indicator raised by the physical layer when delivering a frame to upper layer that gives the signal strength at the time of reception of that frame.

In general, a beacon may be sent if there is a coordinator in the network. It can have, as we said, a slotted transmission with 16 slots at maximum. The distance between beacons is called a superframe. We can also have a combination of slotted and random access, and finally which is the most common case, we can have long silent periods (this should be the normal situation in IEEE 802.15.4 devices), and spaced beacons with any kind of slotted or unslotted random access (Figure 4.4). There may be even a network with no beacons and with no coordinator at all like in the ad hoc IEEE 802.11 mode.

2. Physical Layer

The physical layer is straightforward with a small particularity concerning the frequency of operation; as we can note from the table, different modulations are applied in function of the carrier frequency (Figure 4.5). These many frequency bands restrict the usage of the standard because some of them are not allowed in Europe. The total number of channels is 27 where 16 channels remain in the 2450 MHz band, 10 on 915 MHz and 1 on 868 MHz.

2.1   2450 MHz Physical Layer

It generates 250 Kbps and uses the offset quadrature phase-shift keying (O-QPSK) modulation technique. The data is first mapped onto symbols, 4 bits by 4 bits. Then the symbols are mapped onto 32-bit long chips and

IEEE 802.15.4 and ZigBee_{TM}

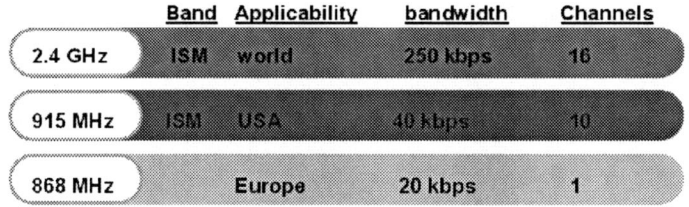

Fig. 4.5. Channel allocation in different countries

Octets: 4	1	1		variable
Preamble	SFD	Frame length (7 bits)	Reserved (1 bit)	PSDU
SHR		PHR		PHY payload

Fig. 4.6. MAC frames

then modulated. Note that in the band (2402–2480 MHz) we can have various peripherals such as IEEE 802.11b/a units, cordless phones, microwave ovens and other equipments. Of course interference can occur between these different wireless units; these aspects should be investigated and quantified. Normally these issues are solved in the 802.19 group.

2.2   868/915 MHz Physical Layer

The data rate is 20 Kbps or 40 Kbps depending on the frequencies (868 and 915, respectively). It uses DSSS and BPSK in the chip modulation and a differential coding for data symbol coding. So data is first differentially encoded, mapped onto chips (15 bits wide) and modulated.

DSSS provides good performance. O-QPSK and BPSK minimize power consumption and reduce complexity.

2.3   PDU Packet Format

The frames that we described before are put into a physical container (Figure 4.6). It starts with a 4 bytes preamble.

The preamble field synchronizes the transceiver with an incoming signal and is set to zero. After the preamble we find the start frame delimiter (SFD) field.

SFD is 8 bits long and it indicates the end of the synchronization, it is composed of a fixed sequence. The length field is for the payload length. If equal to 5 it is an acknowledgement frame and its value should not exceed 127 bytes.

3. MAC Layer

This section describes the MAC where channel access and frame transmission happens. We start by describing the channel access in different possible situations.

3.1 Channel Access

As mentioned, a node may be silent, in a contention free or in a contention access period (CAP). The CAP is delimited by the beacon and the contention-free period (Figure 4.7). It can span between two beacons if there is no contention-free period. This zone is dynamic and may change with time.

All the frames use the slotted CSMA-CA mechanism and transmission should happen completely, i.e. including acknowledgement within the same superframe. MAC layer command messages are transmitted in this zone.

When the system supports beacons, the MAC employs the slotted version of the CSMA-CA algorithm. In the other case, the MAC layer transmits using the unslotted version of the CSMA-CA algorithm. In both cases, the backoff is calculated using exponential backoff periods, where one backoff period is related to the physical layer symbols. The CSMA-CA has been explained before in the 802.11 chapter.

The system uses a congestion window and a variable called *backoff exponent* (BE), which is the number of backoff periods the node waits before accessing the channel again. The delay (congestion window) is increased on busy channel

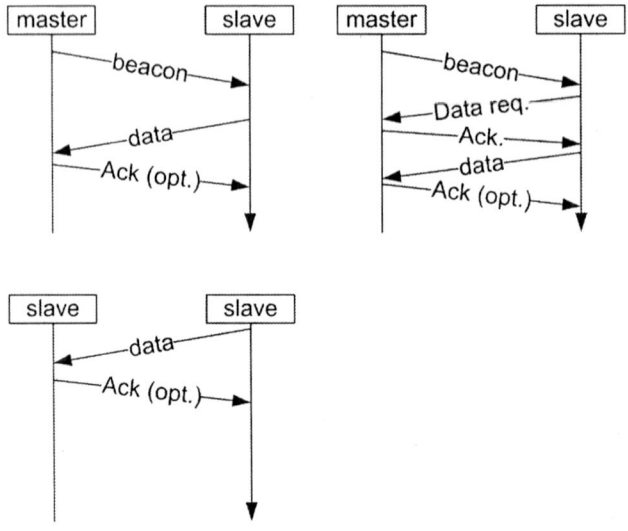

Fig. 4.7. Contention and random access

Command ID.	Name	RFD Transmit	RFD receive
1	Association request	yes	
2	Association response		yes
3	Disassociation notification	yes	yes
4	Data request	yes	
5	PAN ID conflict notification	yes	
6	Orphan Notification	yes	
7	Beacon Request		
8	Coordinator realignment		yes
9	GTS Request		
10	nothing		

Fig. 4.8. Different types of command frames

Bits: 0–2	3	4	5	6	7–9	10–11	12–13	14–15
Frame type	Security enabled	Frame pending	Ack. request	Intra-PAN	Reserved	Dest. addressing mode	Reserved	Source addressing mode

Fig. 4.9. MAC messages for control procedures

and the waited random time before transmission is tossed between 0 and 2^{BE-1}. The congestion window is linearly decreased when the channel is idle.

The second method is contention free. It starts after the CAP and terminates with the next beacon. It is based on grants allocated by the coordinator in a way that the sum of these grants makes the total length of this zone.

Like all other radio techniques there is a need to have some time to process data so transmitted frames are followed by an IFS period, but for acknowledged data the IFS follows the acknowledgement frame.

The types of frames that are transmitted on the physical layer are shown in Figures 4.8 and 4.9.

The orphan notification command is used by any associated node when it looses its synchronization with the coordinator.

A Coordinator realignment command is used when the coordinator receives the orphan notification command or when any PAN attribute changes. It is unicasted to the orphan in the first case and broadcasted otherwise.

The beacon request is related to the scan procedure and is equivalent to a probe request in IEEE 802.11. It is a way to discover local coordinators.

3.2 Energy Detection

During scanning, the full devices measure the peak energy in each channel. This is done by emitting a special frame of this type within a certain interval. This may be repeated over several channels to choose the best one for transmission.

The duration of this ranging process is in number of superframes. At the end the MAC stores the ED parameter for a channel or for all of them for any potential usage.

3.3 Active and Passive Scan

A node that wishes to discover its neighbourhood has the choice of waiting for a beacon or asking explicitly for the presence of coordinators. So it may send an active scan message like the probe request in IEEE 802.11 or it could simply wait for the reception of any beacon on different channels. This scanning is combining with upper layers service discovery accomplished by the ZigBee stack.

3.4 Association Procedure

The association procedure occurs in the presence of a coordinator. The node asks for association and gets in reply a short address such as the piconet number in Bluetooth. The difference here resides in that it is from 0 to 0xffff with the exception of some reserved addresses. The association response is acknowledged by the node.

3.5 Guaranteed Time Slot

This service is dedicated for reservation of bandwidth on the channel. These time slots are only manageable by the coordinator. The guaranteed time slot (GTS) may overlap one more than one superframe so it can be a repetitive system. A reservation mechanism allocates the duration as stated before according to all the PAN needs. In the same way it is allocated by a query from the node to the coordinator, it may be deallocated by any of the two. A device may use a region in the contention-free period while also transmitting in the contention access zone. Since the channel is good for send and receive, i.e. there is no separation with a duplex technique, the device initiating the guaranteed procedure specifies whether it is a transmission or a reception or both at the same time.

To request the allocation of a new GTS the node sends a GTS request with the length and direction subfields. The coordinator acknowledges the message and studies the situation. If there is enough space in the contention free zone, it sends back in the next beacon a GTS structure accepting the query and specifying the index of the slot where transmission or reception is to happen (Figure 4.10).

The GTS request command is used by any associated node that needs a new GTS or to deallocate an existing GTS from the PAN coordinator. Only devices that have a valid short address shall use it.

The GTS length field contains the number of superframe slots requested. The GTS direction subfield is set to 1 for a receive-only GTS and 0 for a transmit-only

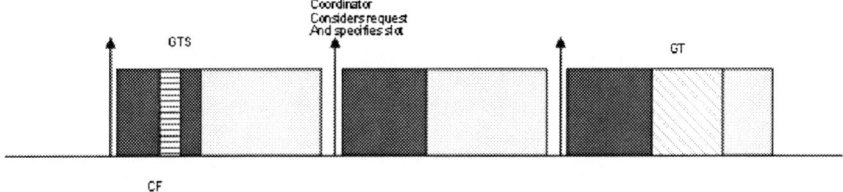

Fig. 4.10. Mix of guaranteed time slot and contention access period

Bits: 0–3	4	5	6–7
GTS Length	GTS direction	Characteristics type	Reserved

Fig. 4.11. GTS Field Structure

Table 4.1. Modulation characteristics of the physical layer

Frequency	Bandwidth	Symbol rate		Data characteristics		
		Chip rate Kchip/s	Modulation	Bit rate Kb/s	Symbol rate ksymbol/s	Symbols
865/915	868–868.6	300	BPSK	20	20	Binary
	902–928	600	BPSK	40	40	Binary
2450	2400–2483.5	2000	O-QPSK	250	62.5	16 orthogonal

where GTS direction is relative to the direction of data frame transmissions by the device.

The characteristics type field is set to 1 if this is an allocation or 0 for a GTS deallocation (Figure 4.11).

4. Security

We will not go into security details for IEEE 802.15.4 but we have to mention that different levels of security are supported in this first version. Table 4.1 gives a short overview of the possible combinations and settings. The security is mainly specified in the ZigBee stack, but relies on encryption and HMAC in the IEEE 802.15.4 MAC layer. Table 4.2 gives the main encryption systems supported in the standard. It also supports integrity and freshness, but the mechanisms are defined in the ZigBee specifications while the encryption is here.

5. Frame Structures

The standard divides its frames into four categories: beacon, data, acknowledgement and control (Figure 4.12). The frames start as usual by a preamble that is not shown here but that is described later.

The frame control field contains information such as:

- Frame type (one of the four types).

Table 4.2. Security modes and their values in ZigBee frames

Method name		Access control	Data encryption	Frame integrity	Segment freshness (optional)	
0 × 03	AES-CCM-64	X	X	X	X	7.6.3
0 × 04	AES-CCM-32	X	X	X	X	7.6.3
0 × 05	AES-CBC-MAC-128	X		X		7.6.4
0 × 06	AES-CBC-MAC-64	X		X		7.6.4
0 × 07	AES-CBC-MAC-32	X		X		7.6.4

Size in bytes 2	1	0/2	0/2 or 8	0/2	0/2 or 8	Variable size	2
Frame control	Sequence number	Destination PAN ID	Destination address	Source PAN ID	Source address	Frame Payload	FCS
		\multicolumn{4}{Address fields}			
\multicolumn{6}{MHR}	MAC payload	MFR

Fig. 4.12. General frame structure

Table 4.3.

Frame type field (first two bits in the frame)	Kind
000	Beacon
001	Data
010	Ack
011	Mac command
100–111	Not used

- Security field: 1 bit indicating whether it is encrypted or not.

- Frame pending field indicating segmentation. In that case the receiver has to send a message asking for the rest of data from the sender.

- ACK request means of course that the receiver should acknowledge the reception.

- Intra-PAN says whether the MAC frame is for the same PAN or to another cluster.

- Address mode says if we are using short addressing or normal MAC addresses for both sender and receiver.

The sequence number is used to number beacons and normal data messages. Any acknowledgement should use that number in its response.

The frame type field is as shown in Table 4.3:

IEEE 802.15.4 and ZigBee~TM~

| Frame control | Sequence number | Addresses | Superframe specs | GTS | Pending address | Beacon payload | FCS |

Fig. 4.13. Different fields in the header control

Bytes 1	0 or 1	variable
GTS specification	GTS directions	GTS List

Fig. 4.14. Structure of the GTS frame

Bytes 0-1	1	4 or 10 bytes	N	2
Frame control	Sequence number	Address fields	data payload	FCS

Fig. 4.15. Data frame structure

5.1 Beacon Frame

The header fields are repeated in the four kinds but, for example, we can see that the beacon frame has the GTS field that allocates slots as explained before (Figures 4.13 and 4.14).

The superframe field has the following control information:

- Beacon order saying at what interval we should find a beacon

- The superframe order gives length of time during in which the superframe is active for receiving.

- Final CAP gives the last slot of the CAP zone

- Battery field is a parameter to advise that the node is not applying the backoff mechanism but sending before the backoff end

- PAN coordinator is set to one when the message emanates from itself and association permit is set to one if the coordinator is accepting new associations

The GTS field is shown in Figure 4.15.

The GTS specification field contains a descriptor for this GTS and a bit indicating that the coordinator is accepting requests.

The GTS directions mask is 7 bits and contains a mask with the directions of the GTSs in the superframe. The lowest bit in the mask is for the direction

Frame control 2 bytes	Sequence number 1 byte	FCS
MHR		MFR

Fig. 4.16. Acknowledgement frame

Bytes 0-1	1	4 or 10 bytes	1	N	2
Frame control	Sequence number	Address fields	Command type	Command payload	FCS

Fig. 4.17. Control frame structure

of the first GTS. One means receive only GTS and 0 is a transmit-only GTS. Of course it is relative to the node sending the request.

GTS list fields are duplicated for each node that has requested GTS and it contains three elements:

- Node's short address
- Start of allocated slot
- Length of allocation

5.2 Data Frame

Acknowledgement frame

This frame is very important. It provides active feedback from receiver to sender (Figure 4.16).

The control frame that is used for usual management procedure in the cluster (Figure 4.17).

The FCS field is 16 bits in length. It contains a 16-bit ITU-T CRC. The FCS is calculated over the MAC Header (MHR), and MAC payload parts of the frame.

6. ZigBee

It is a trademark such as Wi-Fi and WiMAX. Its stack is depicted in Figure 4.18, that only defines some functionalities in layers on top of the IEEE 802.15.4-2003 standard previously defined in order to bring the set of programming tools for the intended market. ZigBee Alliance provides the network layer (NWK) and some applications. The ZigBee Alliance differentiates between a usual node, a coordinator and a router; the difference between the last two is only when the cluster is connected to some external world.

The key definition in ZigBee is called the ZigBee Device Object (ZDO), and addresses three main items:

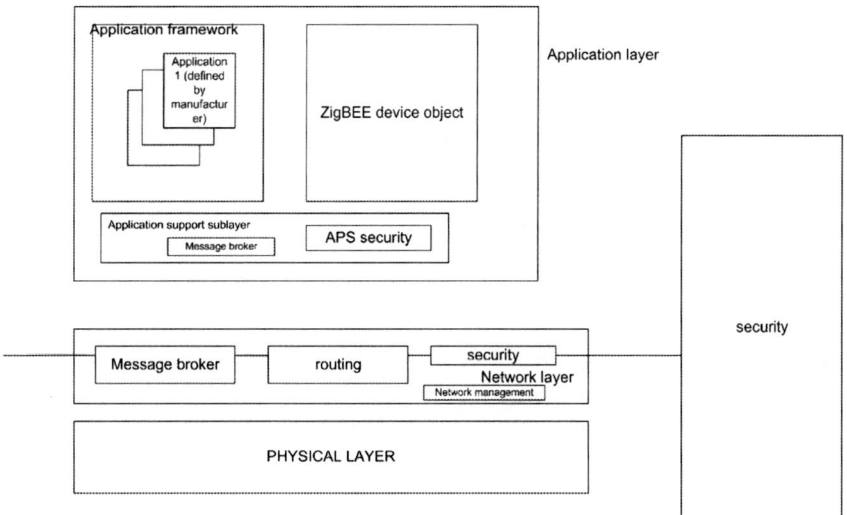

Fig. 4.18. ZigBee general stack

- Service discovery

- Security

- Binding

The role of discovery is to first discover nodes; then they use unicast messages to inquire about the ZigBee Coordinator's/router MAC address, the IEEE address. This may also result in a query about addresses of the rest of associated devices. There is also a broadcast inquiry for the coordinator's/router addresses. There is of course a distinction between the MAC and the network addresses, clearly IP but the field is left open for any further layer 3 courageous proposal.

The equivalent of the Address Resolution Protocol (ARP) is also de facto supported in ZigBee by the same mechanism: a broadcast with the network address or with the MAC address shall return the other missing one.

ZigBee is similar to Bluetooth using profiles.

A profile defines the applications running on top of a device but is more evolved than the Bluetooth where it also specifies the members and the possible actions. So it is a set of device requirements, that collaborate to fulfil a role. An example could be a blood pressure probe, a thermometer and a console together forming a patient monitoring application. We can see that ZigBee is trying to reach more upper level specifications than Bluetooth because it is specified in a time several paradigms such as UPnP, Jini, OSGI are appearing and are proposing similar approaches (without the radio part of course).

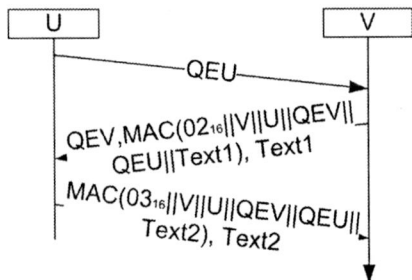

Fig. 4.19. Security handshake

So as profiles play important roles in ZigBee, the discovery is also facilitating the procedure for locating some services via their profile identifiers.

The security services in this object have the role to authenticate and derive the necessary keys for data encryption (Figure 4.19). In that sense, ZigBee complements IEEE 802.15-4 for security.

Key establishment is used as in Bluetooth to authenticate and derive a shared secret, called a link key. An initiator device and a responder have a pre-established trust. Three steps follow this trust establishment: an exchange of ephemeral data, derivation of the link key and a confirmation.

The master secret shared between devices could be sent before with an out-of-band channel. Today it is called a proximity authentication channel and is generalized in several similar protocols.

The network manager is another component of the ZDO. It is implemented in the coordinator and its role is to select an existing PAN to interconnect. It also provides the creation of new PANs. It implements also a routing algorithm between routers of different PANs that it has discovered.

The last component is called the binding manager and has the role to bind nodes to resources and applications but more important to bind devices to channels in order to calculate the remaining available bandwidth a coordinator can grant on the link.

7. Conclusion

This short chapter is a brief introduction to IEEE 802.15.4 or ZigBee. We thought it would be good to have such a low data rate interface compared to other techniques for the sake of having a complete set of technologies covering a large area of applications.

Chapter 5

WiMAX$_{TM}$ and IEEE 802.16

In this chapter we describe the main technical characteristics of the IEEE 802.16 standard. We go through the different layers and their architecture, but at the beginning we explain the main objectives of the standard and some political issues behind it. As it will be seen in the chapter we cannot simply skip physical layer details even though they are not really in the target of this book, because IEEE 802.16 has the very bad characteristic to work in a cross-layer design manner where everything ends up controlling the physical layer.

1. Introduction

The IEEE 802.16 group has started to produce recommendations for a relatively long period. The evolution of the wireless physical layers is seen in the different versions, the same way it can be noticed in IEEE 802.11 standard. That is why we can see a first physical layer implementing plesiochronous digital hierarchy (PDH) like data rates with a line of sight restrictive condition. Few years later, with the familiarization to OFDM, a new version has come up with "line of sight" restriction removed but with lower throughput. We did not see any IEEE 802.16 equipment in the first editions of the standard, not because the lack of products, but because of the unclear legislation in that area together with the wide deployment of fixed asymmetric digital subscriber line (ADSL) wired lines.

Anyway IEEE 802.16, which is a MAN wireless architecture, currently offers two use cases and soon there will be a third means of interconnection. The first two methods are fixed backhaul and cellular like system. The third architecture is mesh interconnection. We will give a short description of the differences in the subsequent sections.

1.1 Backhaul Solutions

The first use case is based on fixed IEEE 802.16 equipment. Independent of the used version, we can employ the technology in a fixed infrastructure as shown in Figure 5.1. In this scenario, the two possible cases are the deployment of point-to-point connections that can span tens of kilometres. The second scenario is the deployment of local loop alternatives to ADSL where the receiver is located inside each house or inside customer premises (Figure 5.2). The latter may use omni or angular antennae.

1.2 Mobile Solutions

The second use case of the standard is based on a different version and provides a mobility scheme for cellular like data transmission. In that case IEEE 802.16 provides the physical and MAC layer only to support the handover. It is different from the 3G family of standards that provide a complete architecture which includes management of the subscribers and the whole network, while IEEE 802.16 stops at layer 2. This also means that handover is done between contiguous base stations and is not provided for interdomain operation.

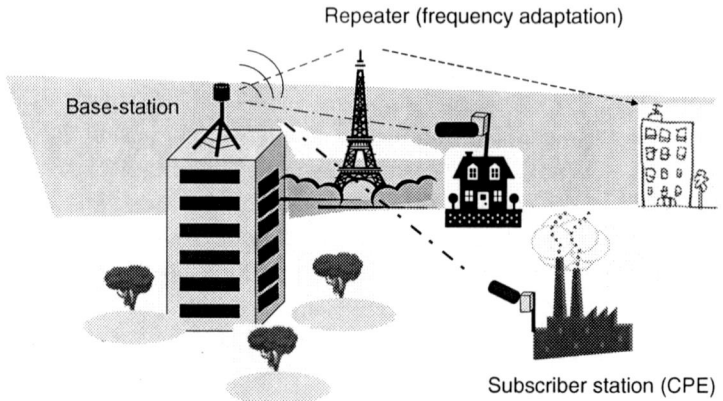

Fig. 5.1. IEEE 802.16 first deployment scenario

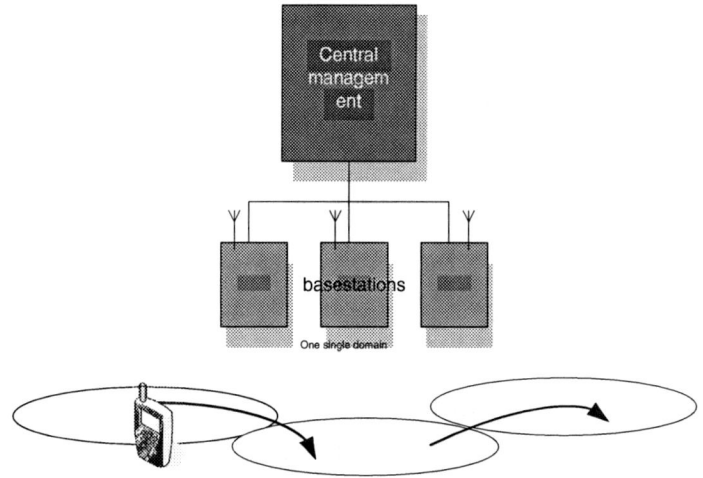

Fig. 5.2. IEEE 802.16 second deployment scenario

1.3 Business Model

The standard is targeted to rural areas in the beginning of development especially when local loops are not available. It is in that case financially much more efficient and effective. Normally, the base stations reside in a central point accessible to the Internet and the outside world by satellite links or any other means. They connect to remote transceivers called customer premises equipments (CPEs) that we will simply call stations in the rest of the chapter. The CPEs are for the time being located on top of the roofs and there should be a second technology that concentrates traffic from the whole building to that CPE. In a later view we would find the CPEs deployed in the residences themselves and hence we would not need anything else to interconnect the final customer to the IEEE 802.16 infrastructure. The usage of this technology is also regulated, since power has to be in the range of few watts and frequency is generally located in the 3.5 GHz band. Usage depends hence on the regulations and it is not planned for the moment to deploy mobile IEEE 802.16 (in Europe) as it is influenced by mobile cellular licences that are usually a national strategic decision dilemma. IEEE 802.16 finds its deployment also in developing countries that do not have the licence problem.

A last word has to be said on differences between IEEE 802.11 and IEEE 802.16. As it is going to be shown, IEEE 802.16 concentrates its protocols on QoS. We can use the ISM bands to operate this technology and hence we may be competing with Wi-Fi.

In fact Wi-Fi and IEEE 802.16 have not really common concepts in between except that they can use the same frequency. IEEE 802.16 is already to some extent an affordable technology, but it is more expensive than Wi-Fi. What we can say is that IEEE 802.16 is a real competitor of IEEE 802.11e with QoS. It is useless today to try to operate IEEE 802.11e on long-distance networks (point to point) and it is much more ingenious to use the QoS of IEEE 802.16. The latter was originally designed for that purpose and brings improvement and advantages over Wi-Fi.

1.4   Evolution of the IEEE 802.16 Group Activities

The group started proposing new drafts a decade ago. It has since then updated the different layers with new techniques. We have the IEEE 802.16 proposal followed by the IEEE 802.16-2004, the IEEE 802.16e mobile amendment and IEEE 802.16j for mesh networks.

1.5   Equipment Category and Frequency Bands

We have already explained the presence of the base station and CPE in the general scenario. There is a third element called "repeater" with the role of relaying circuits and switching frequencies for optimization problems. Practically there might be no difference among the three of them. However, for economic and performance purposes we design a base station with much more processing power, memory, etc. than the CPE or the repeater (Table 5.1).

1.6   Layers and Architecture

IEEE 802.16 is again part of the IEEE 802 group and hence should conform to the bridging (or any layer-2 rules) concepts. The addressing is based on MAC addresses and the base station is seen as a bridge. Nothing prevents it to be a router, but it should implement all layer-2 functionalities also. Addressing, as it is going to be explained, is used as an identification for nodes, but the use of circuits with a circuit identifier replaces this address as long as the node is recognized by the network.

The layers are divided into a MAC and a physical layer. In the MAC layer several sublayers are defined. The first is called the service-specific and is the adaptation of IEEE 802.16 to the available packet types: ATM, Ethernet and IP. Of course, in practice it is only IP that is generalized in the available products.

The MAC sublayer is based on a connection-oriented principle. It is very close to the ATM transport protocol. The connection is using a context that describes the mapping between the incoming flows and the underlying QoS. As we will see later, a station registers itself to the base station, negotiates the physical layer characteristics and then can communicate bidirectionally. A service flow defines the negotiated QoS for all matching packets (service-specific sublayer). The QoS can be changed dynamically and it supports extremely well data bursts. Everything is negotiated for uplink and downlink separately.

Table 5.1. IEEE 802.16 nomenclature

Designation	Applicability	Options	Duplexing alternative
WirelessMAN-SC$_{TM}$	10–66 GHz	–	TDD FDD
WirelessMAN-SCa$_{TM}$	Below 11 GHz licensed bands	AAS ARQ STC mobile	TDD FDD
WirelessMAN-OFDMA$_{TM}$	Below 11 GHz licensed bands	AAS ARQ Mesh STC mobile	TDD
WirelessMAN-OFDMA	Below 11 GHz licensed bands	AAS ARQ HARQ STC mobile	TDD FDD
WirelessHUMAN$_{TM}$	Below 11 GHz license-exempt bands	AAS ARQ Mesh STC	TDD

2. MAC Layer

The MAC layer specifies the operation in point to multipoint and mesh. Point to multipoint is the standard operation and means that a base station is sending data to stations (CPEs). Naturally, since we are in a master–slave radio scenario, the stations hear the base station simultaneously and this is the reason for the choice of that term. The mesh operation concerns relaying and will not be detailed in this chapter. Mesh involves transferring flows from a frequency to another or from a node to another, it is more an optimization problem for network operators.

One difference between point to multipoint and mesh operation resides in the usage of a different parameter to distinguish neighbours in the mesh scenario. It is called link ID and is 8 bits wide. A station or a base station identifies

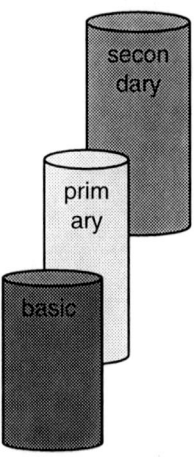

Fig. 5.3. IEEE 802.16 management connections

itself by the MAC address (48 bits unique ID). It is used at the beginning for identification and authentication, but later on it is replaced by a connection ID (CID) (16 bits wide) that will be explained next.

At first, the station establishes its management connections (Figure 5.3). Three types of management connections are defined in the standard (basic, primary, secondary), and each of them will have its CID. They are different in terms of priority and therefore use different QoS parameters.

We will explain hereafter the main mechanisms used by IEEE 802.16 followed by its MAC PDU usage and structure.

2.1 Automatic Repeat Request

ARQ is a selective retransmission at the MAC layer. Some people think it is useless and time consuming as this is solved by the transport layer if necessary (e.g. TCP protocol), and others think it is a must and push in the direction of generalizing ARQ usage everywhere. It is a philosophical problem where each side has good reasons to believe in his opinion. ARQ is optional and applied on a per connection basis. It is either triggered or not, but we cannot alternate ARQ-enabled frames and other types on the same connection. The ARQ message structure is shown later, ARQs go through control messages and can be piggybacked with other data messages.

Data are cut into blocks where a block size has also been negotiated during initialization. ARQ supports a selective acknowledgement system, which means that we can ask for retransmitting some blocks within a received window of data. Negative ACKs is also an option, which means that the receiver did not get the given block. Recall that TCP protocol implements a very limited selective ACK version in one of its extensions (TCP-SACK).

ARQ blocks may be in four states: *not-sent*, *outstanding*, *discarded* and *waiting-for-retransmission*. First, a block is *not-sent*. It is sent and remains *outstanding* with a timeout guard *ack_retry_timeout*. After that, it is either acknowledged or discarded. If a problem occurs, it goes to a *waiting-for-retransmission* state.

At the receiver when a protocol data unit (PDU) is received, its integrity is checked through the checksum. The receiver has a sliding window. When a block with a number that belongs to the window range is received, it is accepted or otherwise rejected. The receiver also maintains a timer per block, which helps to advance the window or to ask for retransmission, then assisting in delivering the good blocks to the upper layers.

An ACK is sent for each block. Blocks outside the sliding window are acknowledged in a cumulative way. Acknowledgements for blocks within the sliding window may be cumulated. Frequency for the acknowledgements is left to each implementer.

2.2 Scheduling and QoS

QoS is an essential brick of the standard. An IEEE 802.16 connection is associated to a specific QoS. Management messages dynamic service addition (DSA) and dynamic service change (DSC) are used for that purpose. Four classes of service are supported: Unsolicited Grant Service (UGS), Real-time Polling Service (rtPS), Non-real-time Polling Service (nrtPS) and Best Effort (BE). Without going into historical fights, the four classes are almost directly inspired from the ATM traffic management classes (CBR, VBR, ABR and UBR)[1].

UGS. UGS supports real-time, fixed size packets at fixed intervals (typically VoIP). It uses Maximum Sustained Traffic Rate, Maximum Latency, Tolerated Jitter and Request/Transmission Policy as parameters. The only unclear parameter here is the last one, which is related to contention policy (described later) allowed in the station.

rtPS. RtPS supports real-time data with variable-sized packets issued at periodic intervals, such as video traffic (MPEG). The parameters are Minimum Reserved Traffic Rate, Maximum Sustained Traffic Rate, Maximum Latency and Request/Transmission Policy.

---

[1] ATM is almost abolished in the operator's networks.

nrtPS. NrtPS is used for non-delay-sensitive applications. They have variable-size packets with a minimum required data rate. The QoS parameters are Minimum Reserved Traffic Rate, Maximum Sustained Traffic Rate, Traffic Priority and Request/Transmission Policy.

BE. Although it is best effort, it still has some parameters, otherwise it would have been too simple...: they are Maximum Sustained Traffic Rate, Traffic Priority and Request/Transmission Policy.

Note that additionally the standard defines ways to "steal" bandwidth from other connections, to piggyback traffic or queries and to poll for additional bandwidth requests.

Generally, connections are negotiated through the three control channels described at the beginning. QoS is assigned by polling stations or by explicit requests sent to the base station.

Figure 5.4 shows the way physical slots are assigned by the base station at maximum interval times.

Similarly, Figure 5.5 below shows minimal allocation in an FDD scenario (the terms maximum and minimum times represents the relative time slot that the base station can refer to for bandwidth reservation).

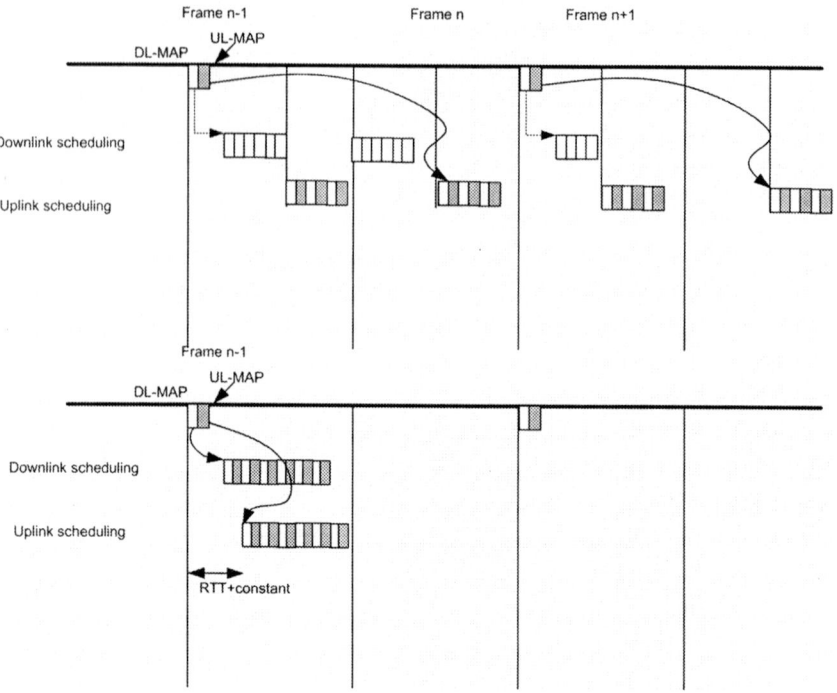

Fig. 5.4. IEEE 802.16 scheduling with TDD scheme

Name	Frequency range	Features	Scheduling
Wireless MAC SC	10-66 Ghz	-	TDD/FDD
Wireless MAN SCa	Below 11 Ghz, license	AAS,ARQ,STC,Mobile	TDD/FDD
Wireless MAN OFDM	Below 11 Ghz, license	AAS,ARQ,MESH, STC,Mobile	TDD/FDD
Wireless MAN OFDMA	Below 11 Ghz, license	AAS,ARQ, HARQ,MESH, STC,Mobile	TDD/FDD
Wireless Human	Below 11, unlicensed	AAS,ARQ, HARQ,MESH, STC,Mobile	TDD

Fig. 5.5. IEEE 802.16 scheduling with FDD scheme

The allocated slots for the Uplink MAP (UL-MAP) messages, containing information that defines the entire access for a scheduling interval, are those after the round trip delay period.

2.3 Contention Resolution in IEEE 802.16

Normally, if we use allocated bandwidth, contention will not happen. However, collisions may happen during "Initial Ranging" (ranging is the mechanism that enables the station to decide which physical layer parameters to use for optimal data transmission) and "Request" messages, since they occur on unallocated bandwidth. Contention is a problem in uplink channels to the base station and not on the downlink naturally. One mandatory contention resolution algorithm must be implemented and is called the *truncated binary exponential backoff window*. The window is controlled by the base station that broadcasts its parameters. When a station wants to send data and **even if there is no collision detected**, it sets its backoff window equal to the value found in the UL-MAP message. The station selects a random number within the backoff window that will indicate the number of contention transmission opportunities the station waits before transmitting. The transmission opportunities in a contention zone are also set by the base station in the UL-MAP messages.

After transmission, the station waits for the acknowledgement. In case it is not received, the station increases its backoff window by a factor of 2 up to the maximum backoff window. The station then uses a new random number within its new backoff window and repeats the same procedure. After a maximum retry number the station has to try ranging again to see what to do in the uplink channel to avoid such problems.

Transmission Opportunities in IEEE 802.16. The transmission opportunity is a period in minislots to allow stations to transmit their bandwidth requests. The term has nothing to do with IEEE 802.11E. It is intended for a group of stations to transmit their bandwidth requests or Initial Ranging exchanges in those periods, which make up in fact the contention zone. The duration of these opportunities depends on the parameter published by the base station and on the physical layer characteristics.

2.4 Adaptive Antenna System

Adaptive antenna system (AAS) is a system using directive antennae or antenna elements that adapt the antenna beam for a better reception. In general, the beam is better focalized onto the receiver to increase the spatial density of receivers without increasing the transmission rate or frequency. In IEEE 802.16 beams are steered to multiple users simultaneously and realize frequency reuse within the cell, a reuse that is proportional to the number of antenna elements.

Another possible physical improvement happens in the signal-to-noise ratio (SNR) gain if the receiver combines multiple signals by adjusting the gain to a particular emitter. Another general advantage of this technique is the reduction of interference by steering null beams for co-channel interferers. AAS is used in IEEE 802.16 and its negotiation requires a set of substantial messages between the base station and the AAS-enabled stations.

2.5 Joining an IEEE 802.16 Network and Initialization

As it can be seen, IEEE 802.16 is very complex and a lot of parameters need to be agreed upon before any lucky transmission. Network entry and initialization is the term used in the standard to define the necessary procedures for entering and registering a new station. It is defined for PMP and mesh, with special additions for the mobile case.

The procedure is summarized as follows:

- Scanning for downlink channel and establishing synchronization with the base station
- Obtaining base station parameters
- Performing ranging procedures
- Negotiating basic capabilities
- Authenticating and key derivation
- Registering the station
- IP initialization
- Setting up connections

These procedures are primarily implemented through downlink map (DL-MAP), uplink channel descriptor (UCD) and downlink channel descriptor

(DCD) messages that will be briefly described in sections 5.2.6.1.1–5.2.6.1.3 and in the physical layer part.

2.6 PDUs

The MAC PDU format is given in Figure 5.6.

Two kinds of MAC headers are used: the first concerns all generic operations, while the second is dedicated to bandwidth requests (Table 5.2).

The generic header is showed in Figure 5.7.

We have:

- HT: header type, either zero or one to distinguish between generic and bandwidth request frames

- EC: if we use encrypted payload or not

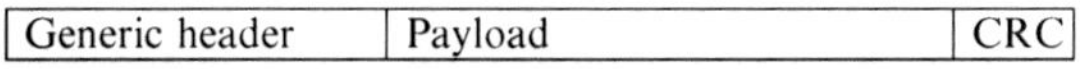

| Generic header | Payload | CRC |

Fig. 5.6. IEEE 802.16 general MAC PDU

Table 5.2. IEEE 802.16 MAC PDU header subtypes

Code for header type	Name of command	Explanation
000	BR incremental	Additional data requested in bytes
001	BR aggregate	New request in bytes
010	PHY channel report	Sent uplink to give the power level
011	BR with UL Tx Power Report	Combination of the two messages
100	Bandwidth request and CINR report	BR and CINR (carrier to interface noise ration) that indicates measured ratio by the MS from the BS. It shall be interpreted as a single value from -16.0 dB–47 dB
101	BR with UL sleep control	Combines BR and power saving modes for the station
110	SN Report	Sequence Number for selective ack. control
111	CQICH allocation request	Channel quality information channel is a separate channel for quality monitoring

HT	EC	Type	RSV	CI	EKS	RSV	Length
Length			CID				
CID			HCS				

Fig. 5.7. IEEE 802.16 generic MAC PDU header

HT	EC	Type	Bandwidth request MSB (11bits)
BR LSB			CID
CID			HCS

Fig. 5.8. IEEE 802.16 bandwidth request MAC PDU header (bit widths in parentheses)

Table 5.3. Available bit rates with different modulations in the SC physical layer

Channel size	Symbol rate	Bit rate for QPSK	Bit rate for 16 QAM	64 QAM	Frame duration	Physical slots per frame
20 Mhz	16	32	64	96	1 ms	4000
25	20	40	80	120	1	5000
28	22.4	44.8	89.6	134.4	1	5600

- Type: different header subtypes (mesh, ARQ, extended header, grant management for piggybacking bandwidth requests, etc.)
- CI: States if we use CRC or not
- EKS: relative to security keying
- LEN: is the packet total length in bytes (including MAC header and CRC if present)

The bandwidth request header type is shown in Figure 5.8.
The length of the bandwidth request header is 6 bytes (48 bits).

- CID refers to an uplink connection
- BR indicates the number of bytes requested
- HCS is the header check sequence

MAC Management Frames. This is the heart of the IEEE 802.16 standard in general. The format is a type of message followed by the payload. The list of messages is given in Table 5.3. We will discuss the main goals of most of them.

From Table 5.3 we can see that the different message types are not always using the same kind of connections, some use the primary channel while others use the basic one. In general, the construction of the message includes a list of elements called "information elements" as structures that uniquely describe a specific parameter. We give hereafter some details on the most important message types.

Downlink Channel Descriptor Message. It is sent periodically to define the physical layer parameters for downlink channel. The different options have to be agreed upon by using this message type and its reciprocal on the *uplink* side. The message contains an important structure called the *burst profile*, since all IEEE 802.16 data is sent as bursts having a coding, modulation length, etc.

Uplink Channel Descriptor Message. An UCD is sent by the base station (the message name could be misleading) at a periodic interval and defines the characteristics of a given uplink channel as seen by the base station, concerning one particular station or a subset of stations.

Downlink Map Message. It is the most important scheduling message since it provides the downlink access and organization of the frame. A station receives this message from the base station and is able to know where in the frame it can find its data. Thus, DL-MAP contains pointers to bursts in the subframe, where each burst may be destined to a different station. The message is detailed in section 3 hereafter.

Uplink Map Message. The UL-MAP message allocates access to the uplink connections. It is sent by the base station. The message contains information elements stating how a given station is allowed to send its data on a specific channel.

Ranging Request. It is sent by the station at transmission initialization and then on a regular basis to measure the delay. In general, ranging is a procedure for determining what kind of physical layer is best adapted for the existing conditions. The ranging message is immediately answered by the base station. When OFDM is used the ranging can be targeted to one or several OFDM subcarriers.

Registration Request. It is sent by the station at initialization. It contains authentication information and some upper layers parameters such as:
- IP version
- Station capabilities encoding
- Vendor-specific information
- Convergence sublayer parameters
- ARQ parameters

Table 5.4. MAC management frame types

Type	Message name	Message description	Connection
0	UCD	Uplink Channel Descriptor	Broadcast
1	DCD	Downlink Channel Descriptor	Broadcast
2	DL	MAP Downlink Access Definition Broadcast	DL
3	UL-MAP	Uplink Access Definition	Broadcast
4	RNG REQ	Ranging Request	Initial Ranging or Basic
5	RNG-RSP	Ranging Response	Initial Ranging or Basic
6	REG-REQ	Registration Request	Primary Management
7	REG-RSP	Registration Response	Primary Management
8	Reserved		
9	PKM-REQ	Privacy Key Management Request	Primary Management
10	PKM-RSP	Privacy Key Management Response	Primary Management
11	DSA-REQ	Dynamic Service Addition Request	Primary Management
12	DSA-RSP	Dynamic Service Addition Response	Primary Management
13	DSA-ACK	Dynamic Service Addition Acknowledge	Primary Management
14	DSC-REQ	Dynamic Service Change Request	Primary Management
15	DSC-RSP	Dynamic Service Change Response	Primary Management
16	DSC-ACK	Dynamic Service Change Acknowledge	Primary Management
17	DSD-REQ	Dynamic Service Deletion Request	Primary Management
18	DSD-RSP	Dynamic Service Deletion Response	Primary Management

19		Reserved	
20		Reserved	
21	MCA-REQ	Multicast Assignment Request	Primary Management
22	MCA-RSP	Multicast Assignment Response	Primary Management
23	DBPC-REQ	Downlink Burst Profile Change Request	Basic
24	DBPC-RSP	Downlink Burst Profile Change Response	Basic
25	RES-CMD	Reset Command	Basic
26	SBC-REQ	Basic Capability Request	Basic SS Basic
27	SBC-RSP	Basic SS Basic Capability Response	Basic
28	CLK-CMP	SS network clock comparison	Broadcast Broadcast
29	DREG-CMD	DE/Reregister Command	Basic
30	DSX-RVD	DSx Received Message	Primary Management
31	TFTP-CPLT	Config File TFTP Complete Message	Primary Management
32	TFTP-RSP	Config File TFTP Complete Response	Primary Management
33	ARQ-Feedback	Standalone ARQ Feedback	Basic
34	ARQ-Discard	ARQ Discard message	Basic
35	ARQ-Reset	ARQ Reset message	Basic
36	REP-REQ	Channel measurement Report Request	Basic
37	REP-RSP	Channel measurement Report Response	Basic
38	FPC	Fast Power control	Broadcast
39	MSH-NCFG	Mesh Network Configuration	Broadcast
40	MSH-NENT	Mesh Network Entry	Basic

Table 5.4. Continued

Type	Message name	Message description	Connection
41	MSH-DSCH	Mesh Distributed Schedule	Broadcast
42	MSH-CSCH	Mesh Centralized Schedule	Broadcast
43	MSH-CSCF	Mesh Centralized Schedule Configuration	Broadcast
44	AAS-FBCK-REQ	AAS Feedback Request	Basic
45	AAS-FBCK-REP	AAS Feedback Response	Basic
46	AAS_Beam_Select	AAS Beam Select message	Basic
47	AAS_BEAM_REQ	AAS Beam Request message	Basic
48	AAS_BEAM_RSP	AAS Beam Response message	Basic
49	DREG-REQ	SS Deregistration message	Basic

Privacy Key Management Messages. Privacy key management (PKM) messages is a series of requests and responses to authenticate the station to the base station and to derive a session key. More details will be given in the chapter dedicated to wireless security protocols.

Dynamic Service Addition Messages. The QoS parameters related to GS, RT services and so on are negotiated through the DSA-Req/DSA-Response messages.

ARQ, Mesh and AAS. Other specific messages are used to negotiate the automatic repeat request mechanism used for loss recovery. As explained before, ARQ helps recovering lost bursts.

A set of messages is reserved to configure mesh communication between stations, but we do not go deeper here on this topic.

AAS is the intelligent antenna system (AAS). These messages are exchanged between AAS-enabled stations and their base station to agree upon specific antenna beams they prefer to communicate upon. During ranging the station detects base station available beams, then negotiating which one to use and its physical parameters.

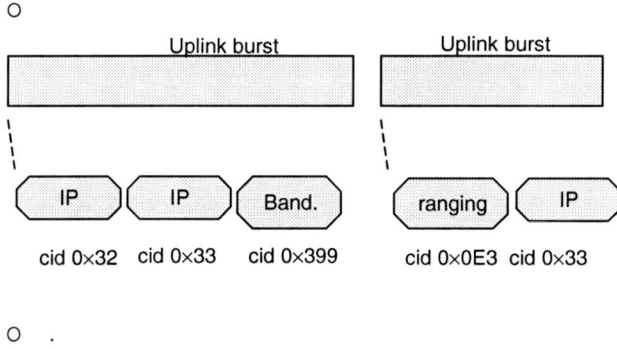

Fig. 5.9. Typical bidirectional transmission with ranging

2.7 PDU Organization

The IEEE 802.16 standard is capable of doing concatenation, fragmentation and packing (Figure 5.9).

- Concatenation means to send different data corresponding to different CIDs between a station and the base station in a single burst or consecutive ones.

- Fragmentation means that we cut an upper layer data unit (SDU) in smaller pieces and send them numbered.

- Packing means gathering multiple MAC SDUs into a one MAC PDU.

- The appropriate connection attributes say if we have fixed or variable length packets.

3. Physical Layer

The physical layer is a heritage of several evolutions that are called SC, SCa, OFDM, orthogonal frequency division multiple access (OFDMA) and Mobility addendum (802.16E), as it can be seen in Table 5.1. Additional physical layers are being worked out in the committees. It operates in a large band of 40 GHz, which leads to many possibilities. The scheduling of connections can happen in time, frequency and space domains. This means that we use time division duplex (TDD) to share uplink and downlink, but also frequency division duplex (FDD), OFDMA and directive antennae that restrict their beams to a specific end point. The space domain is also managed through multiantenna techniques (MIMO) that appeared in the late editions of the standard.

The physical layer is very complex because it can merely combine all the above-mentioned mechanisms into one transmission event. It is also strongly based on the MAC signalling layer, the MAP messages that were described previously. The role of unimax forum is to agree on subset of these parameters for a smooth transition.

3.1 IEEE 802.16 SC (Single Carrier)

The first physical layer is for upper frequencies greater than 11 GHz with a single carrier. One big advance in IEEE 802.16 is that each frame is coded and modulated dynamically according to channel quality. This is similar to IEEE 802.11. Framing is an important concept in IEEE 802.16 and helps in synchronization. A frame is used to send and receive data at one side. For example, the TDD mode mandates that a frame be divided into a downlink subframe followed by an uplink one. In FDD of course they are multiplexed on different frequencies and can hence be emitted at the same time.

Frame duration can be of 0.5, 1 or 2 ms. We can send broadcast half-duplex or full-duplex subframes. Some transitions may happen with no transmission at all or transmission may be continuous.

In Time division mode there are two mandatory periods that are left empty (no transmission from both sides) to have time to (de)activate the necessary radio frequency equipment from transmission to reception and vice versa. The smallest unit that can be specified is different in the uplink and downlink subframes. It is called a "physical slot" (PS) in downlink and "minislot" in uplink. A minislot is equal to 2^m PS, where m ranges from 0 to 7.

The Downlink subframe is the part of the frame going from the base station to the station. It is composed of a preamble as usual (coded in QPSK) followed by a data structure containing pointers to data bursts which follows in the subframe. The data pointers point to data that may have different encodings.

The downlink units listen to this header and find the beginning of each burst where they will find another header containing the destination MAC address. If it matches, the downlink unit address is picked and decoded, otherwise it is dropped.

The FDD downlink (Figure 5.10) is similar in principle but contains also a part for half-duplex stations where the scheduling for subframe (n) is programmed in subframe ($n - 1$) preceding it. The DL-MAP message is specific for each burst. It contains several pieces of information such as CID, modulation, FEC and most important the start position of the burst itself in the subframe. The size of the burst in terms of PSs is also calculated here.

A burst is hence divided into a certain number of physical slots (Figure 5.11 and Figure 5.12). This number varies for a fixed size burst because the modulation is not always the same and varies according to the signal quality. Then the FEC is applied on the PS and it generates a code word. The upper layer is

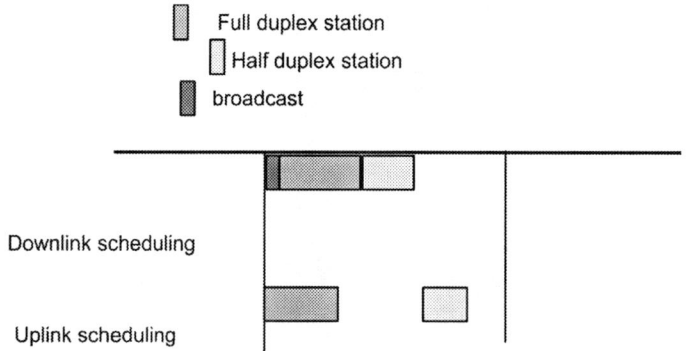

Fig. 5.10. IEEE 802.16 FDD operation

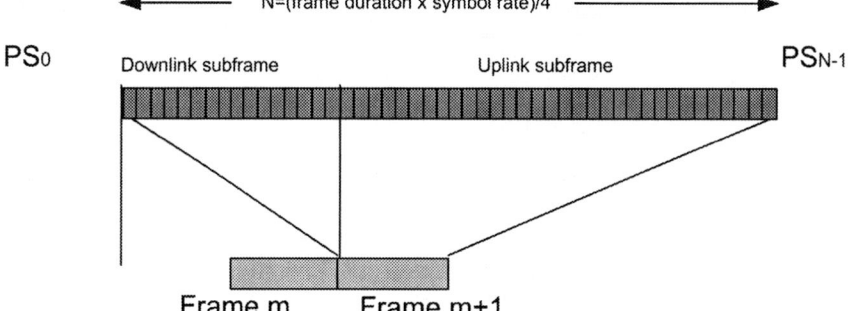

Fig. 5.11. Structure of the TDD frames

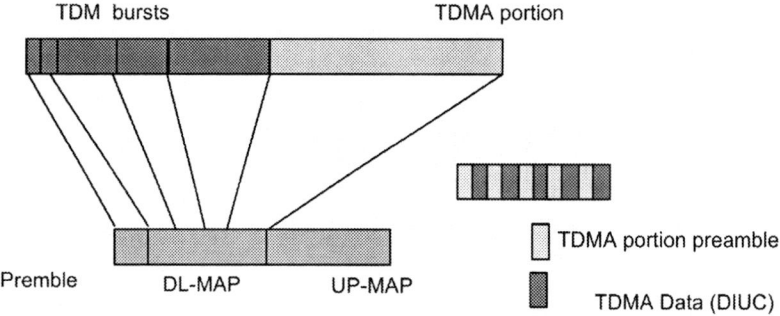

Fig. 5.12. IEEE 802.16 TDD operation

responsible for cutting the layer-2 frame into code words that are subsequently cut into smaller pieces. Data coming from the upper layers follows the sequence:

- Randomization (a mechanism that helps for clock recovery at the receiver)
- FEC

- Preamble preparation

- Symbol mapping

- Baseband pulse coding

- Modulation

FEC.   FEC can be one of four kinds, mainly based on Reed Solomon codes. The first three are based on Reed Solomon, whereas the fourth is a Turbo Code type.

Code type 1 is Reed Solomon only with variable protection. It is recommended for fast data rates coding. The second is Reed Solomon as well, but with convolution codes. It is recommended when data rates are slower. It requires a soft decoder and gives improved signal protection. The third is also Reed Solomon and adds parity check. It is recommended for moderate to high data rates. Parity is used for data check and correction. The last is a turbo code type and is used when carrier to interference ratio is bad.

Downlink Modulation.   As we have previously indicated the modulation is dynamic and adaptable. The modulation techniques yet supported are QPSK, 16-QAM and 64-QAM. A burst is the unit that specifies a single modulation technique. After measuring the quality, the transmitter may change modulation scheme or change transmission power.

A pulse shaping is applied to the signals (I and Q) to cut excess bandwidth by means of a specific transfer function. The two signals are transmitted to the antenna after multiplying each of them with the carrier frequency.

Uplink.   Three kinds of bursts are allowed in the uplink. The first is in the contention zone for establishing a first connection. Ranging, that means measuring and adapting to transmitter/receiver best parameters, uses this kind of profiles.

The second is also a contention burst, but happens after a polling or a broadcast from the base station.

The third burst type is the classical data transmission case over contention free zones already allocated by a reservation mechanism.

Bursts are structured like in the downlink case and have a time guard called subscriber station transition gap (SSTG) that helps the receiver to distinguish between bursts coming from different stations. Several kinds of messages circulate in the contention zone like ranging messages previously explained and bandwidth request messages.

A burst has a preamble and a control structure called the UL-MAP that contains the CID relative to the connection, the message code (e.g. Ranging,

Data grant, Request, End of MAP) and the offset indicating the beginning of the information in minislots.

Each burst is characterized by a modulation, an FEC and a randomization initialized by a seed.

The lower procedures (modulation, FEC and randomization) are the same as in downlink operation.

Bandwidth in the SC. Despite the huge available band, regulation in every country mandates the baseband to be reduced to a reasonable amount. The table hereafter gives the numbers for the SC technique and the corresponding physical slots in the frames.

As it can be seen we can reach a good bandwidth of 134 Mbps with the 64-QAM scheme under a channel size of 28 MHz.

3.2 The Single Carrier in Lower Bands

Similar to the first SC physical layer, the SCa (Figure 5.13) option also supports TDD and FDD operation. It has additional FEC methods to ensure robust transmission but also has a no-FEC mechanism that is generally combined with the automatic repeat request option (ARQ). A STC option is available and tuning for adaptive antennae is added (AAS).

The 256-QAM and BPSK schemes are added to the physical layer. Data is coded and modulated on a burst basis. The burst is structured by a burst set preamble and a ramp down period to give time to distant equipment to switch to other receivers. Pilot words are regularly sent to the other end for synchronization purposes. So from these bulletts we can see how close the two systems are.

Space Time Coding (MIMO and AAS). IEEE 802.16 SCa supports a STC mechanism with two antennas. The basic idea is that two blocks are correlated and sent to two transmit antennae. At the receiver the two blocks are received on both receiver antennas and checked for errors.

Figure 5.14 gives a very coarse idea on the mechanism called "paired block" transmission/reception. The burst set consists of a preamble and a repetition of n bursts in the STC. Since this technique is different from the normal time

Channel size	Symbol rate	Bit rate for QPSK	Bit rate for 16 QAM	64 QAM	Frame duration	Physical slots per frame
20 Mhz	16	32	64	96	1 ms	4000
25	20	40	80	120	1	5000
28	22.4	44.8	89.6	134.4	1	5600

Fig. 5.13. Available bit rates with different modulations in the SC physical layer

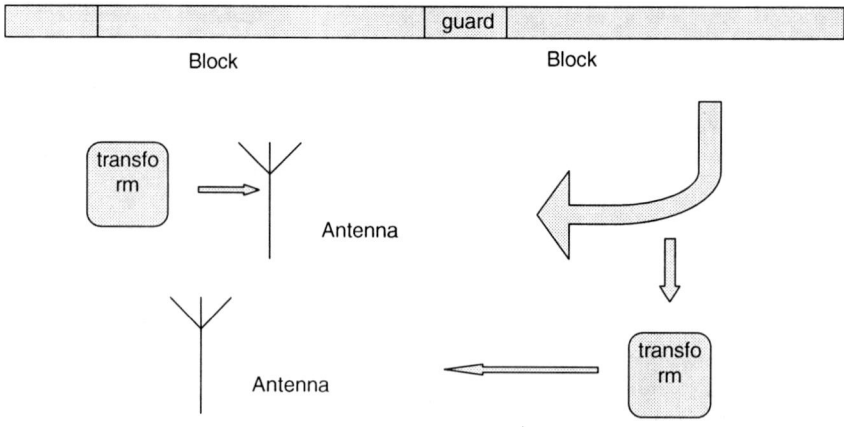

Fig. 5.14. Paired blocks sent on multiple antenna

division method explained before, it cannot be sent simultaneously. A burst set may hence be STC coded or not, but it may not contain the two kinds at the same time for decoding reasons.

Subchannel Frame Transmission. When several subchannels are used, there is a method to cut the data and reassemble it at the receiver.

- The procedure first adds a base frame preamble to the payload.

- The data is then cut into small blocks. Each consecutive block is sent over a subchannel.

- We form an appended block by adding a unique word at the end of the block.

- Create repeat segments for FEC.

- Take each segment (a group of repeated appended blocks) and form a burst set of segments by adding two parameters at the beginning of each segment set.

- Concatenate these sets of segments to form a burst and transmit over the subchannel.

AAS Operation. AAS helps in improving spectrum utilization by optimizing the transmission through multi-element antennae and intelligent antennae. The range is better, the SNR ratio is improved, interference is reduced and frequency reutilization is possible. AAS requires that both transmitter and receiver be aware of this capability.

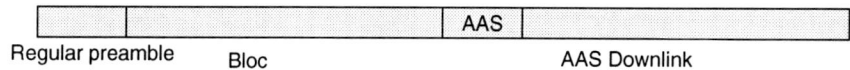

Fig. 5.15. IEEE 802.16 OFDM with TDD operation in AAS mode

Procedures for requesting bandwidth and for ranging are slightly different due to the fact that the base station may be aiming its beam to another part and hence the bandwidth message may be lost.

Note that AAS uplink operation can be in a situation where several stations send simultaneous bursts to the base station without interference and this is the main reason why we have a specific signaling for this kind of operation. In that case, if we assume a TDD mode where bursts are scheduled to different downlink receivers with some of them supporting AAS and others do not, a very complex mapping header would be required. The format of messages shall be with the non-AAS information first, followed by the AAS enabled bursts (Figure 5.15). The partition between the two parts in a single frame is described in the frame header, called the Frame control header (FCH) explained next.

Frame Control Header. The first burst in the downlink contains the FCH. It is composed of a DL-MAP message, an Uplink one, a DCD that describes the physical layer of the channel, and one or many Uplink channel descriptors UCD indicating the physical characteristics of uplink channels. Any station should understand the FCH and shall be able to send its own FCH. The main contents of an FCH are:

- Preamble length

- Pilot Word Interval

- Power adjustment, that states if the sender is transmitting all the burst at peak or average power

- Modulation type of the FCH payload

- FEC code

- DCD burst profile

MAP Messages. The MAP messages belong to the MAC but relate to the physical layer. They help the control of the physical layer. A number of MAP messages are defined such as:

- Power control to command the change of the station transmission power.

- Concurrent transmission, meaning that the bursts that are going to be sent are scheduled in parallel.

- AAS mark for uplink traffic, used to inform the base station that it is now receiving AAS bursts.

- UL-MAP that informs that data is sent in an uplink channel or on different subchannels for a certain connection. We put in the case of subchannels in the overall list of channels.

- Burst set delimiter extended MAP message helps to indicate the end of a burst set and the beginning of a new one, which may have different characteristics.

3.3 OFDM Physical Layer

The OFDM option is provided for low-frequency operation (below 11 GHz) in a non-line-of-sight scenario.

OFDM is based on orthogonal symbols and uses inverse Fourier transform. A symbol time structure is shown in Figure 5.16, being the concatenation of a useful symbol time T_b for a given data to be sent and the last T_g, which is the period used to collect multipath measurements to maintain orthogonality between signals. This collect period (CP) is used by the transmitter and is kept always fixed. The role of the receiver at the beginning is to guess this duration among a set of values and once it is found, the receiver is said to be synchronized in the time domain.

In the frequency domain the OFDM symbol is based on subcarriers. The number of subcarriers is determined by the Fourier matrix size. There are three kinds of subcarriers:

- Data subcarriers

- Pilot subcarriers used for measurement

- Null subcarriers to keep guard bands and synchronize

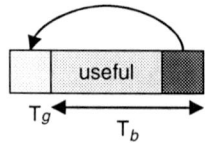

Fig. 5.16. Symbol time structure for OFDM

The IEEE 802.16 OFDM is parameterized by four primitives:

- The bandwidth of the channel; it is multiple of 2.0, 2.75, 1.75 or 1.5 MHz
- The number of subchannels; it is equal to 200 and the FFT size is 256
- The sampling factor, that gives the distance between subchannels and hence the duration of a symbol, equal to 8/7 in general
- The ratio between the CP and the useful time, 1/4, 1/8, 1/16 or 1/32

Channel Coding. The same steps are applied as in the SC and SCa physical specifications. A randomization followed by FEC and interleaving are used.

FEC is mainly using Reed Solomon method and optionally two variants of Turbo codes.

Interleaving happens by two permutations. One permutation spreads adjacent coded bits on different subcarriers and the second permutation permutes adjacent bits onto less and more significant bits of a constellation to spread possible errors.

Modulation, Preamble and Frame Structure. OFDM physical specification supports QPSK, 16 QAM, 64 QAM. The downlink is adaptive and works on an allocation basis as explained before. The uplink is less open in order to support low complexity stations.

The preamble in OFDM is based on the transmission of 128 samples with the CPs as shown in Figure 5.17.

The preamble is sent in downlink subchannels in a specific way too complicated to explain and beyond our scope. Uplink preamble is only 64 samples long and is sent over even subcarriers.

The frame base for OFDM transmission is sent in FDD or TDD (Figure 5.15). A frame is divided into downlink and uplink subframes.

The uplink is used for ranging, bandwidth request and for PDU transmission from different stations. The downlink consists of a single PDU. This PDU has its own preamble followed by the FCH. The DL-MAP messages sent in bursts are following the FCH. The uplink subframe follows with upper link MAP messages that are based on a contention mechanism with backoff.

The transmission of bursts happens in an integer value of OFDM symbols, which means that filling is necessary before transmitting to the encoder. Note that the AAS is also supported in this OFDM specification. A base station can

Fig. 5.17. Preamble structure for OFDM

hence alternate directed beam with preambles and bursts for specific stations and diversity generalized transmission for the rest of the receivers.

Ranging and Space Time Coding. Two kinds of ranging are supported in the physical layer: the first occurs on the first connection and the second is periodic.

The first ranging may require several periods and exchanges until the final parameters are accepted by the base station. This initial ranging can happen on OFDM subchannels as well as on AAS supported systems.

Finally we note that OFDM can be combined with STC with two antennae.

3.4 OFDMA Physical Layer

The last physical layer specified in 802.16 is the OFDMA. It is based on OFDM operating in low bands below 11 GHz and in non-line-of-sight conditions or mobile conditions if we adopt the IEEE 802.11E addendum.

OFDMA is similar to OFDM in terms of symbols and procedures. Subcarriers are however put together in groups, and each group is or may be allocated to a specific connection. This means that a group of subcarriers called a subchannel may be allocated in the downlink to a receiver and in uplink it could be reserved to a given station. Figure 5.18 shows subcarriers grouped into three subchannels.

The primitives explained in the OFDMA physical layer are the same, i.e.

- Bandwidth

- Number of subcarriers, FFT is 2048 instead of 256

- Sampling factor

- G ratio for the CP (ratio of CP time to useful time).

A symbol structure in OFDMA is hence composed of data pilots and zero subcarriers. The composition of this number of subcarriers is negotiated through MAP, but follows a minimum set of calculated rules.

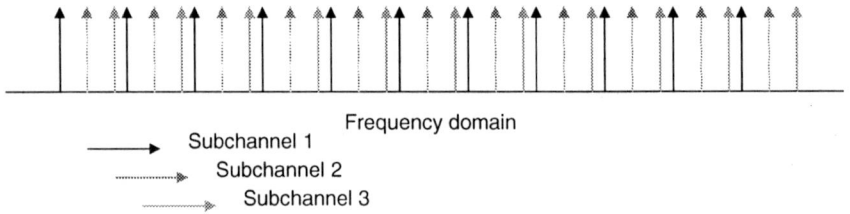

Fig. 5.18. OFDMA with 3 subchannels

*WiMAX*_{TM} *and IEEE 802.16* 149

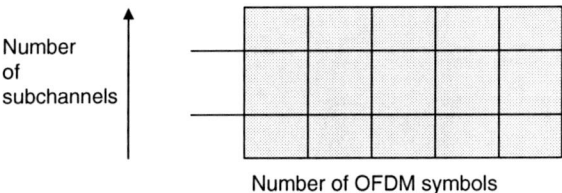

Fig. 5.19. A typical "region" allocated for a packet in OFDMA transmission

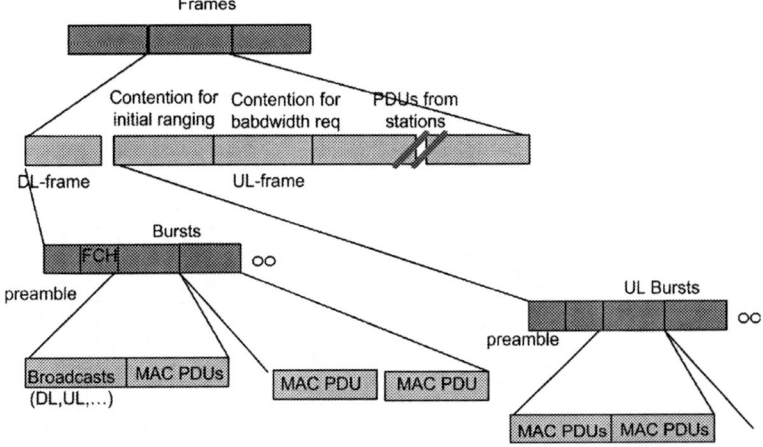

Fig. 5.20. Hierarchical structure of Data for Downlink and Uplink modes

Slots in OFDMA.   A slot in OFDMA is defined in time and in subchannels:

- In downlink full usage of subchannels mode, we have one slot on one subchannel using one OFDMA symbol.

- In partial usage of subchannels one slot uses two OFDMA symbols and one subchannel.

- In uplink partial usage of subcarriers, one slot is one subchannel and three symbols.

The picture here shows the definition of an OFDMA data region.

The term *segment* is a group of subchannels used per MAC connection. The matrix in Figure 5.19 shows how a MAC frame is segmented into OFDMA slots. First, it is cut into slots and then mapped onto subchannels that are allocated to that connection. In IEEE 802.16E the subchannels are divided into six groups (Figure 5.20).

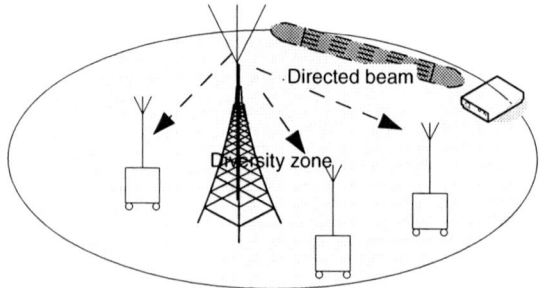

Fig. 5.21. 802.16E directed beam

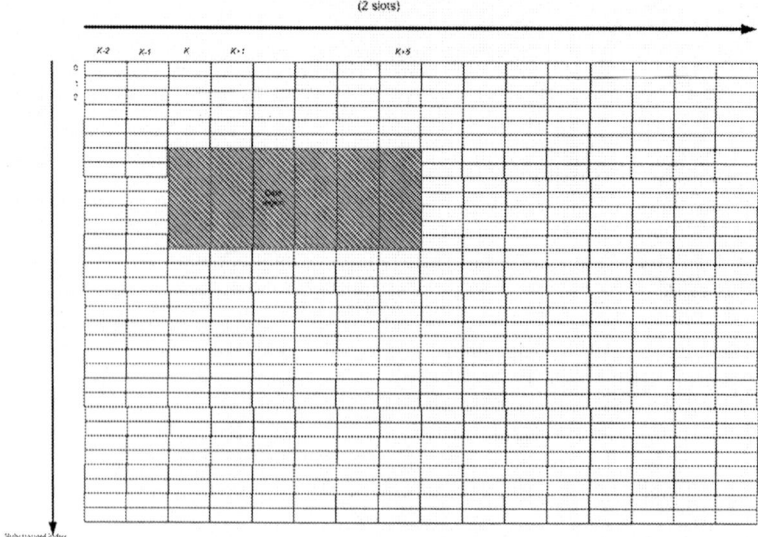

Fig. 5.22. A Mac Frame mapped onto OFDMA slots

Few differences in IEEE 802.16E show that, for example, the way slots are mapped onto subchannels are different. Here we show only the uplink channel for OFDMA IEEE 802.16E (Figure 5.21).

Again like in previous modes we support TDD or FDD but not both. The frame structure is as usual started by a preamble followed by the downlink period and the uplink period (see Figure 5.23). Time guards are inserted to give time for the hardware to switch on and off. Figure 5.22 shows the mapping of these periods onto OFDMA subchannels.

OFDMA also supports AAS mechanisms where the non-AAS region of the frame is sent before the AAS-enabled burst. When using the AAS technique a dedicated link is reserved for all measurement, ranging, bandwidth request, MAP messages. Figure 5.24 gives a simple block diagram of 802.16E transmitter.

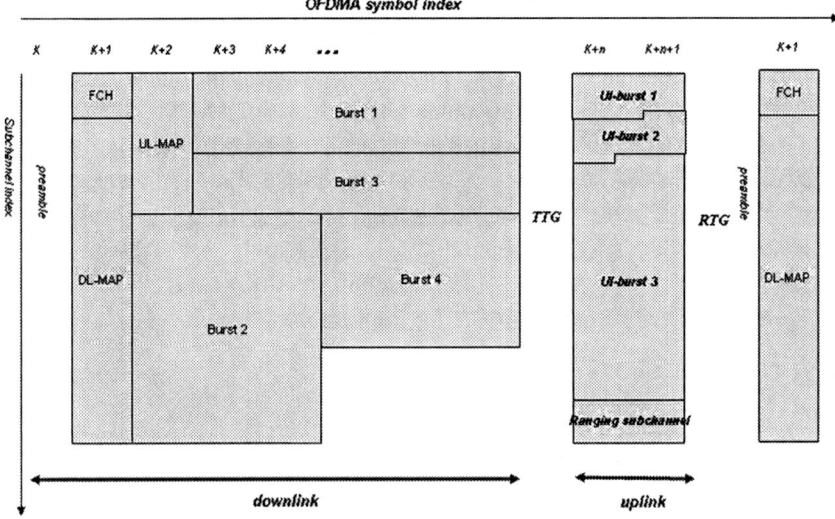

Fig. 5.23. Partitioning Frames in the two direction for OFDMA

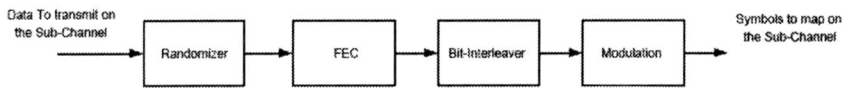

Fig. 5.24. Simple block diagram for 802.16E physical layer

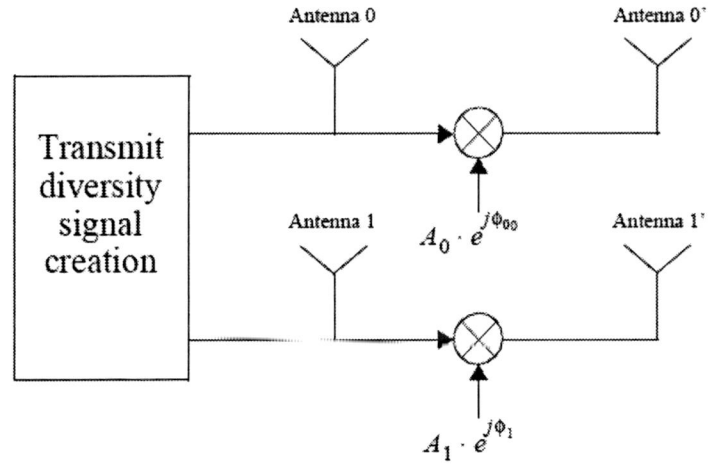

Fig. 5.25. MIMO frames mission with 4 antennas

Frames are from 2 to 20 ms long in both TDD and FDD. The MAP messages used in the physical layer context are for defining the burst size (number of symbols, subchannels, etc.), the AAS parameters (such as number of antennae). One new MAP message supported in this physical layer is the MIMO MAP that specifies some basic MIMO parameters. The STC here support four antennae instead of two in previous physical versions. Among the MAP messages for physical layer configuration we have the messages that define how subcarriers are shared among the used subchannels.

OFDMA uses convolutional coding and optional turbo coding. Figure 5.25 shows an example of transmission using four antennae.

Chapter 6

Security in WLAN, WPAN, WSN and WMAN through Wi-Fi$_{TM}$, Bluetooth$_{TM}$, ZigBee$_{TM}$ and WiMAX$_{TM}$

Security constitutes a crucial problem for wireless networks whose diffusing of transmission is shared. Whereas Wi-Fi/Bluetooth products flood the market on a worldwide scale and give access to Internet, the companies become increasingly demanding security guarantees. Security thus seems one of the major stakes for the operators. Additionally, securing the link does not secure the complete transmission and requires complex authentication procedures that establish a trust between entities that are not logically related.

To meet this challenge, large efforts are made by the concerned actors (industrial, scientific community, operators, etc.). Many work is undertaken within the framework of consortia of industrials and organizations of standardization in order to bring robust solutions in terms of security. Among these authorities, we can mention the IEEE working groups such as 802.11i, 802.1X, 802.15.1/4 and 802.16.

Communications can take place only if the communicating entities are in the same radio range. The passive listening of communications on a channel is completely obvious being given the broadcasting nature of the medium. However, the use of direct antennas (through AAU and MIMO) can increase considerably the extent of the covering zone and allow a foreign entity far away from the network to listen to and inject traffic in the system.

Among the main attacks listed in the context of wireless networks, we quote:

- Intrusion: the intrusion consists, for an external element, to connect itself to the operator access point and then to be able to penetrate in the network (WLAN, WPAN, ...). This attack can occur in two manners, either coming from the interior of the network or remotely.

- Data capture: it is the most traditional attack. It is easy to recover the contents of data which circulate on the wireless link. This can take place thanks to spies entities which are in the zone or are remote with the help of a direct antenna.

- "Man-in-the-Middle" attack: man-in-the-middle (MITM) attacks on a LAN are the most frightening attacks. It acts, for a machine having bad intentions, to make so that the communication between two target computers passes from the assailing machine and this in a transparent way. This type of attack is extremely powerful, because the attacker can hear all the communications and modify what he wants. This attack is particularly easy to implement, it is enough to have an access point like Trojan horse.

- Intruding access point: this attack consists in connecting a false access point to the network. After its installation, the attack can be done near the access point, in its radio range or remotely by using a direct antenna.

As to make life easy for attackers, several tools are available on the free open source domain that gather all the mechanisms ready to be used, one can mention as an example the *Auditor* bundle.

As for fixed networks, security mechanisms (encryption, authentication) must be implemented in a wireless network. Nevertheless, because of the context of wireless communication, the wireless link is very vulnerable, it is important to be vigilant with respect to the malicious attacks which benefit from the radio wave propagation in open space. The radio channel is a broadcasting medium, that propagates in all directions and thus is interceptable by whoever provided with a suitable device.

This chapter aims to review the mechanisms proposed for WLAN, WPAN, WMAN and WSN networks, residual faults, the progress of standardization, as well as the implemented solutions. We restrict ourselves particularly to the four systems, Wi-Fi, Bluetooth, WiMAX and ZigBee. This chapter is composed thus of four main parts. The first part presents a state of the art of security in the Wi-Fi systems. It includes a presentation of the basic mechanisms, an analysis of the security faults and a classification of the listed possible attacks. This part ends with a description of the solutions suggested, a progress report of the activity of standardization and also a description of the problems which remain open. The second part is devoted to the Bluetooth system, covering the same aspects as in the case of the Wi-Fi system. The third and fourth parts address, respectively the security mechanisms related to WiMAX and ZigBee systems.

A general conclusion ends this chapter by giving hot lines to set up short- and long-term security solutions.

1. Security in Wi-Fi Systems

Security in wireless networks gained an increasingly important interest since the publications of scientific articles which clearly describe the insufficiencies of the used mechanisms; particularly those concerning the IEEE 802.11/Wi-Fi systems. Confidentiality and authentication must be taken into account since most wireless architectures are connected to a fixed infrastructure.

1.1 Security in Wi-Fi Networks

Wi-Fi products, present on the market, mainly support security mechanisms proposed by IEEE 802.11 standards (802.11b/802.11i/WPA/WPA2) with some owners improvements.

Basic Mechanisms. The IEEE 802.11 standard defines basic mechanisms to enhance security. These mechanisms are:

- Access control

- Authentication of stations which are connected to the network

- Confidentiality and integrity of data thanks to the WEP protocol.

These mechanisms are limited to make secure exchanges between two stations.

Unfortunately, these techniques cause many faults generating major security problems in IEEE 802.11 networks. The first articles which sounded alarm on these large gaps were published in 1999 and were written by researchers of the University of Berkeley.

In what follows, we will review each one of these mechanisms.

Access Control. Access control is performed mainly using two techniques based on the use of SSIDs and/or access control lists (ACLs).

- SSID: the SSID is an identifier of the network represented by an alphanumeric code common to all the access points and stations forming the same network. Thus, if a user knows the SSID of the network, then it is able to reach it if not, its request will be rejected. Default values exist in the solutions suggested by the manufacturers and must be personalized by the administrators of these networks. This protection is weak because the access point periodically sends the SSID in clear in beacons frames which can be easily intercepted;

– ACL: the second technique of access control consists in using lists of access. Indeed, each access point has a list where are registered all the MAC addresses of authorized stations. The access point carries out a filtering on the basis of MAC addresses. Each list must be continuously updated, manually or by a specialized software, in order to add or to remove users. This protection is also weak because only a passive listening of the network is sufficient to recover the MAC addresses.

Be added to these two techniques, the use of passwords. These three means of protection can be used in a complementary way.

Data Confidentiality. This function requires the exchange or pre-shared encryption/decryption keys. The mechanism of exchange of keys is not specified by the standard but depends on the manufacturer's choice. The data ciphering is carried out using the WEP algorithm. We focus in the following part on the description of the operation of WEP as well as its properties.

Encryption. The standard IEEE 802.11 proposes a traditional data encryption technique based on using a symmetric key (secret) (Figure 6.1). Data protection is carried out thanks to the use of the encryption algorithm proposed by WEP which is done according to the diagram of Figure 6.2.

The message to be encrypted (called plaintext) undergoes a transformation using a function parameterized by a key. The result of the encryption algorithm is called ciphertext and can then be transmitted. An intruder can then listen to or completely copy the ciphertext. However, this intruder is supposed not to have the key of deciphering, what makes impossible the recovery of information sent in clear.

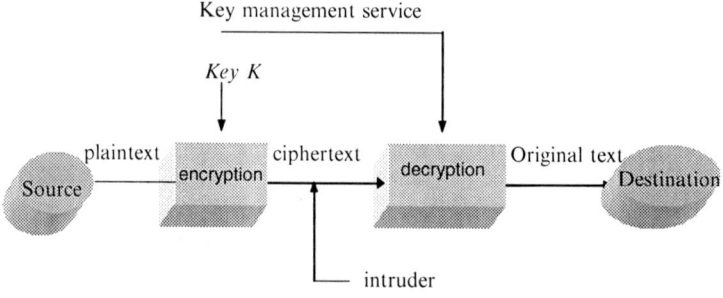

Fig. 6.1. Data encryption/decryption

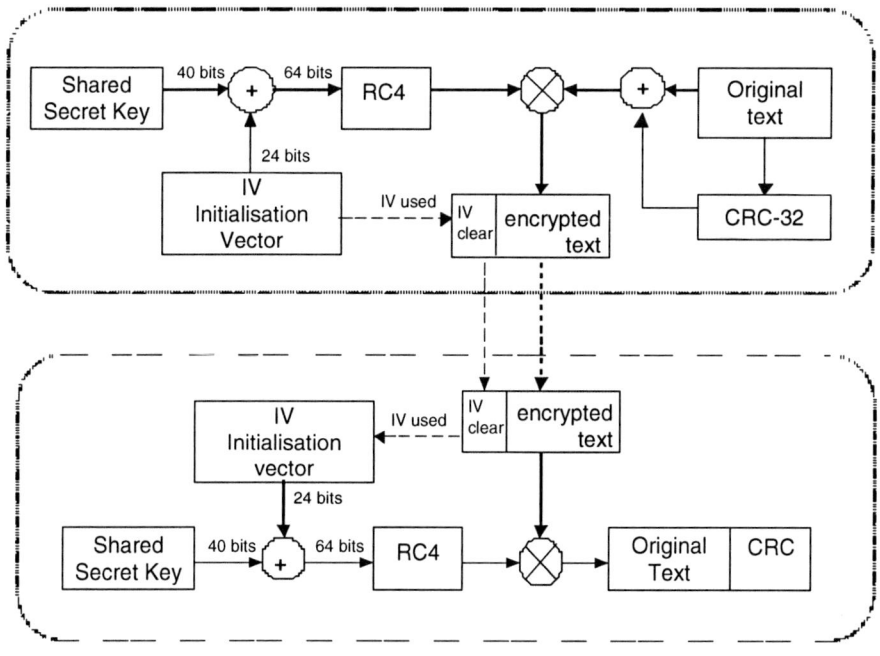

Fig. 6.2. Data encryption/decryption using WEP

WEP is based on an algorithm with symmetrical secret key, the same keys are thus used for the encryption and decryption. Encryption is applied on the cleartext sequence and on a pseudo-random sequence. The opposite process is applied during decryption. To enhance security, in addition to the static encryption key present in each station, an initialization vector which changes with each sent frame is used. Key is called keystream which is in fact obtained by applying the RC4 algorithm [FLU 01] to the secret key k and the initialization vector (IV), noted RC4 (IV,k).

Data encryption by WEP, illustrated in Figure 6.2, is done through the following steps:

- The two entities share a secret 40 bits key k. This key is concatened with an IV (24 bits), which is reset at each transmission. IV offers the property of autosynchronization of the algorithm.

- The resulted 64 bits are used as an entry for the generator of pseudo-random sequences, which uses RC4 algorithm, to produce the encryption key sequence.

- In parallel, the integrity algorithm CRC-32 is applied on the text and produces a value of 32 bits ICV in order to check the integrity of the transmitted useful data.

- The text to be encrypted and ICV are concatened and are XORed with the encryption key. This constitutes the ciphertext.

- The ciphertext as well as the IV are then sent on the wireless link.

The WEP encrypted frame has the following format (see Figure 6.3):

- An IV sent in clear which is used for the generation of the encryption/decryption keys. It is on three bytes.

- The following byte allows to recognize the key thanks to two bits (Key ID field).

- A MAC service data unit (MSDU) which corresponds to the MAC data frame protected by WEP with size lower than 2304 bytes.

- An encrypted 4 bytes ICV field that allows the checking of the authenticity and the integrity of the frame.

Decryption. The decryption algorithm, illustrated in Figure 6.2, includes the following steps:

- Generation of the decryption key from the IV of the received message and the identifier of the decryption key.

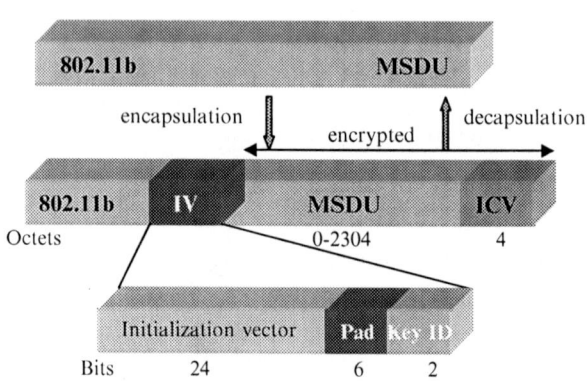

Fig. 6.3. Encrypted packet format
Integrity check value (ICV) Initialization vector (IV) for each frame

- Decryption of the received message with the key to obtain the initial message.

- Check of the integrity of the decrypted message by using ICV algorithm.

- Comparison between the obtained ICV and transmitted ICV with the message; if an error is detected an indication of error is sent to the MAC layer and then transmitted to the transmitting station.

Authentication Models. IEEE 802.11 standard proposes two authentication systems:

- The Open System Authentication (OSA). It is the first technique suggested by default. There is no explicit authentication; a station can join any access point and listen to data in clear within a cell. This mode is used if one chooses an easy use of the system without security management;

- The shared key authentication (SKA).

The second technique provides a better system for authentication based on the use of a mechanism to share secret key between the station and the access point. This authentication is possible only when the WEP encryption option is activated. It allows to check that the station which wishes to be authenticated has the shared key. This operation does not require to transmit in clear the key, but is based on the encryption algorithm described previously.

The shared secret key is stored in the Management Information Base (MIB) in writing mode only. The IEEE 802.11 standard does not give any detail on the procedure of key distribution. 802.11b.corl corrects the problems involved in the MIB and the management of the access points in the IEEE 802.11b version.

This mechanism comprises generally four steps:

- The station wanting to connect to an access point (infrastructure mode) or another station (ad hoc mode), sends an authentication request.

- The access point answers by sending an authentication frame containing a random challenge text of 40 or 128 bits (the manufacturers extended the initial length to reinforce security), generated by WEP algorithm.

- Upon receiving the frame containing the text, the station encapsulates it in an authentication frame and encrypts it with the shared secret key before sending it to the access point.

- The access point decrypts the text using the same shared secret key and compares it with the sent one. If the text is identical, the access point confirms its authenticity otherwise it sends a negative response.

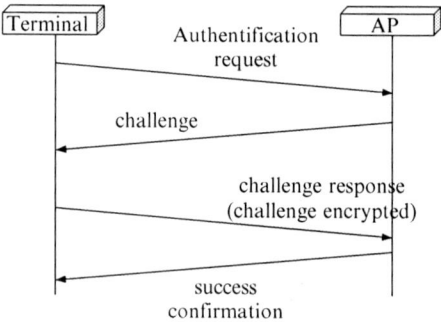

Fig. 6.4. SKA model

We can note many weaknesses in the mechanisms listed below, among which:

- Data encryption protects only MAC data frames, and not the header of the physical layer frame, other stations can then listen the encrypted frames.
- The same shared key is used for encryption/decryption and authentication.
- There is also a method where the stations and the access points may use WEP alone without authentication. That is carried out in open system mode.

Security mechanisms proposed in IEEE 802.11 are included in all the derived standards including IEEE 802.11b (Wi-Fi).

Key Management. No mechanism of distribution of keys is mentioned in the standard. Consequently, the equipment suppliers implement specific solutions in their products. Unfortunately, in the majority of the cases, it is not possible to verify the consistency of the implemented procedures; in spite of that security faults are always detected.

WEP keys can be used in two manners:

- The first method provides a window with four keys (taking into account the reservation of 2 bits for their description in Key ID of the IV (see Figure 6.3). Stations and access points can decrypt frames using one of these four keys. Classically, only one key is used, it is introduced manually and is regarded as being the key by default.
- The second method is called key mapping table. In this method, each MAC address can have a separate key. This table at least contains ten entries and the maximum size is directly related to the chipset. Assign

a separated key to each user permits to better protect the frames. Since keys can be only changed manually, the periodic validity of the life of the key remains an important problem. This solution consumes memory, is difficult to manage, and is error prone. For these reasons, it is often not implemented by many vendors.

1.2 Security Flaws

In spite of the deployed security mechanisms in IEEE 802.11 networks, the security of the latter is a critical problem. This is due to:

- Weaknesses detected in the access control mechanisms
- Problems of insecurity related to WEP

In what follows, we will describe in detail the main detected flaws at the beginning. For more information on these last, the reader can read the articles of Walker, Goldberg and Wagner of the University of Berkeley who were at the origin of the demonstration of the vulnerability of IEEE 802.11 systems; what besides involved a great polemic [BOR 01, WAL 00].

Several weaknesses have been discovered in this protocol compromising security in WLANs. WEP is based on RC4 algorithm to generate the pseudo-random encryption keys from the IV and the shared secret keys. To obtain the encrypted text, an operation XOR is carried out, between the encryption key and the plaintext. At reception, the access point (or a station in an IBSS) applies an operation XOR to the encryption key which it calculates and with the received encrypted text in order to find the original text.

Encryption such as it is used is vulnerable to the following attacks:

- If the attacker changes a bit in the encrypted text, then the equivalent bit in the plaintext will be faded.

- If an attacker intercepts two texts encrypted with the same key, then it will be easy to find the plaintexts. Indeed, by applying the XOR operation on the two encrypted texts, the effect of encryption disappears: the result is then the XOR of two plaintexts. By using statistical attacks, the two texts can then be found.

More explicitly, this can be expressed mathematically in the following way: We assume:

- $T1, T2$ two plaintexts
- k encryption key
- $C1, C2$ two ciphertexts

We have C1 = T1 XOR k and C2 = T2 XOR k
From where
C1 XOR C2 = (T1 XOR k) XOR (T2 XOR k) = (T1 XOR T2) XOR (k XOR k)
= T1 XOR T2

However, WEP protocol supports some solutions against these attacks by the means of:

- Data integrity check thanks to ICV field.

- Change of the IV so that different encryption keys are generated. With each frame, a new IV is created. IV generation is not specified in IEEE 802.11 standard, but some implementations use incrementation by one, or random IV with each frame.

But these two protection techniques are rather weak, were incorrectly used by WEP, what led to serious security faults. These last are related to the components of the global security system including IVs, CRC field, RC4 algorithm, SSIDs and/or ACLs, key management and access points management.

The objective of this part is to explain in details the failure of the initial listed mechanisms in order to better understand thereafter the new orientations to improve them and the existing solutions.

Key Problems. Beyond the WEP insecurity, insecurity of IEEE 802.11 networks is also due to the problems of management of the secret keys. IEEE 802.11 standard does not define any management mechanisms and distribution of these keys. However, security of these networks is closely related to those critical function. Some questions have to be asked:

- How are keys generated?

- How are they distributed?

- How much keys are deployed on the network?

- With which frequency are they changed?

These various points are not specified in the standard and are left to the choice of the manufacturers. Consequently, we raise the following points:

- The majority of keys (40 or 104 bits length) are generated manually. They are then not very different from a traditional password and are vulnerable to dictionary attacks. There is another method for the generation of keys which consists in making correspondence between a secret key and the MAC address. But, this solution is not highly deployed.

- The standard specifies that a protected method must be employed for the distribution of the keys. The method of distribution of keys affects the frequency of change of these keys insofar as if the keys are changed manually, it would be reasonable not to frequently change them.

The reuse of a single key increases obviously the probability of IV collisions which grows proportionally according to the number of the users. Moreover, PCMCIA cards reset IV to 0 each time they reset:

- There is another weakness which refers to management keys in WLANs. Indeed, the constitution of the keys is based on the identity of the machine (MAC addresses) and not on the identity of the user. This can be regarded as a weakness insofar as an unauthorized user can be introduced into the network using an authenticated machine.

- Size of the key, fixed at the beginning, to 40 bits was extended to 128 bits (only 104 bits are actually used). Increase the size of the key does not secure against the repeated accesses but makes nevertheless the attacks more difficult to implement.

IV Problems. Initially, the IV is coded into 24 bits and is transmitted in the message in clear. WEP concatenates IV with the shared keys. As described previously, for each transmission of a frame, a key of a family of 2^{24} keys is selected to encrypt data. This process involves the following weaknesses:

- WEP allows to share the same key between the stations of the same BSS, a key which is not changed. With the help of a distribution of adequate generation of frames one could manage to find the same key. Consequently, it would be necessary to add an algorithm which makes it possible to prevent that a station uses an IV used by another station.

- If one considers a loaded access point which continuously sends frames of 1.500 bytes at 11 Mbps, the space of IV will be completely traversed after $1.500*8/(11*10^6) *2^{24} = \sim 18,302$ s or 5 hours 5 minutes (this value can even be lower because the size of the frames is generally lower than 1.500 bytes). This permits an attacker to recover at the end of 5 hours two encrypted texts.

- Moreover, some implementations authorize that the stations and the access points of the same network use the same secret keys, what increases, chances of collision of the encryption keys. It is recommended to change IV per frame.

CRC Problems. Integrity control field CRC-32, applied only to the useful part of the data, involves primarily two faults:

- CRC-32 is linear, what implies that starting from two texts whose CRC are known, it is possible to find the CRC of the sum of these two texts. This characteristic was intelligently exploited by the attackers; it is possible to change bits in the encrypted message, without changing the integrity of the whole message. In addition, let us note that the CRC permits to check that the integrity of data and not their authenticity. An example is clarified below.

We assume:

- M the clear message, C the encrypted message, C' the modified encrypted message.

- b the changes which the attacker wants to apply to the encrypted text C.

- c is the encrypted ICV.

- c' it is the value of ICV modified by the attacker.

Then $b = C$ XOR C'

The text modified by the attacker has the format $<C', c'>$. Let us search c'.

$<C', c'> = <M', CRC(M')>$ XOR $RC4(v,k)$
$= <M$ XOR $b, CRC(M$ XOR $b)>$ XOR $RC4(v,k)$
$= <M$ XOR $b, CRC(M)$ XOR $CRC(b)>$ XOR $RC4(v,k)$
$= <M, CRC(M)>$ XOR $RC4(v,k)$ XOR $<b, CRC(b)>$
$= <C, c>$ XOR $<b, CRC(b)>$
$= <C$ XOR b, c XOR $CRC(b)>$

The attacker has only to do $c' = c$ XOR $CRC(b)$

- CRC is a non-encrypted function. An attacker can thus create checksums of frames where the plaintext is known. As soon as the attacker obtains a pair (IV, key), he will be able to inject its own intruding packets in the network.

- Access points can accept packets even if they reuse the same IV. An attacker can benefit from the second weakness of the CRC field, quoted above, and to send false packets which will not be rejected by the access point.

RC4 Problems. RC4 algorithm constitutes the critical element of the WEP protocol. RC4 was developed in 1987 by Ron Rivest for RSA Data Security Inc. RC4 aims at generating a series of random numbers, an operation of XOR is carried out between the generated bits and the message of origin. RC4, has the following properties:

- Encryption is carried out bit by bit (flow encryption) without awaiting the complete reception of data.

- The encryption keystream k is independent of the plaintext, the determination of the decryption key k' according to k requires the factorization of an integer N of big size.

- Encryption/decryption are fast, approximately ten times faster than with a DES algorithm.

- The implementation is simple.

- Immunity with respect to a linear or differential cryptanalysis.

RC4 typically operates as follows: RC4 uses a clear field (e.g. IV) and a secret value (e.g. key secret) to produce a pseudo-random sequence (e.g. encryption key). On the other hand, if the same secret value is used with various clear fields, then the analysis of the pseudo-random sequences can help to find the secret field.

Because of the use of RC4 algorithm for the generation of the encryption keys, WEP has a weakness which was at the base of the attack of Fluhrer, Mantin and Shamir. This attack resulted in breaking easily and quickly WEP keys. Several tools available on Internet allow a uer, not informed, to obtain certain RC4 keys (IV, k) known as weak keys. In order to counter this vulnerability, some equipments remove these keys, however, a simple statistical analysis is enough to discover them.

Management Problems. Many access points support an SNMP agent to carry out the management tasks. If the configuration of access points is badly carried out, an intruder can read and potentially write significant information. Moreover, it is possible to reach the configuration of certain access points by an Internet interface (via HTTP or telnet). It is thus important to instal a password to reach the configuration of the access points.

Some information on connections to the access point are stored on the terminals of the users. If the configuration of the user is not optimal, accessing this information becomes very easy and weakens all the system. For example, in the products 3Com, the non-encrypted WEP key is contained in a Windows register.

Exemple: Monitoring Private Data. Encrypted packets can be captured by *LinkFerret (a monitoring tool)*, but the message content appears unreadable. The only information extracted is the message header; recall that this is never encrypted:

No. 133
Time 13.879958
Source D-Link_cd:1c:fe
Destination Aironet_35:be:d6
Protocol Info IEEE 802.11 Data, SN=20, FN=0
IEEE 802.11
    Type/Subtype: Data (32)
    Frame Control: 0x4108 (Normal)
        Version: 0
        Type: Data frame (2)
        Subtype: 0
        Flags: 0x41
            DS status: Frame from STA to DS via an AP
            .... .0.. = No More Fragments
            .... 0... = Retry: Frame is not being retransmitted
            ...0 .... = PWR MGT: STA will stay up
            ..0. .... = More Data: No data buffered
            .1.. .... = **Protected flag: Data is protected**
            0... .... = Order flag: Not strictly ordered
    Duration: 117
    BSS Id: Aironet_35:be:d6 (00:40:96:35:be:d6)
    Source address: D-Link_cd:1c:fe (00:0f:3d:cd:1c:fe)
    Destination address: Aironet_35:be:d6 (00:40:96:35:be:d6)
    Fragment number: 0
    Sequence number: 20
    WEP parameters
        Initialization Vector: 0xeef91d
        Key Index: 0
        WEP ICV: 0xf5461597
Data (56 bytes)
    38 8c 0b 19 f5 7e 6c d4 55 86 f3 27 15 bb 30 ae
    d2 29 9d 42 9a a9 87 3e 83 ac e8 93 17 9d 06 51
    87 56 0f cb e7 55 30 df 1c 16 dc 86 67 2e 8a 54
    05 22 77 2d f5 46 15 97

As previously mentioned, the IV is given in the packet header – this must be clearly transmitted since it changes for every packet.

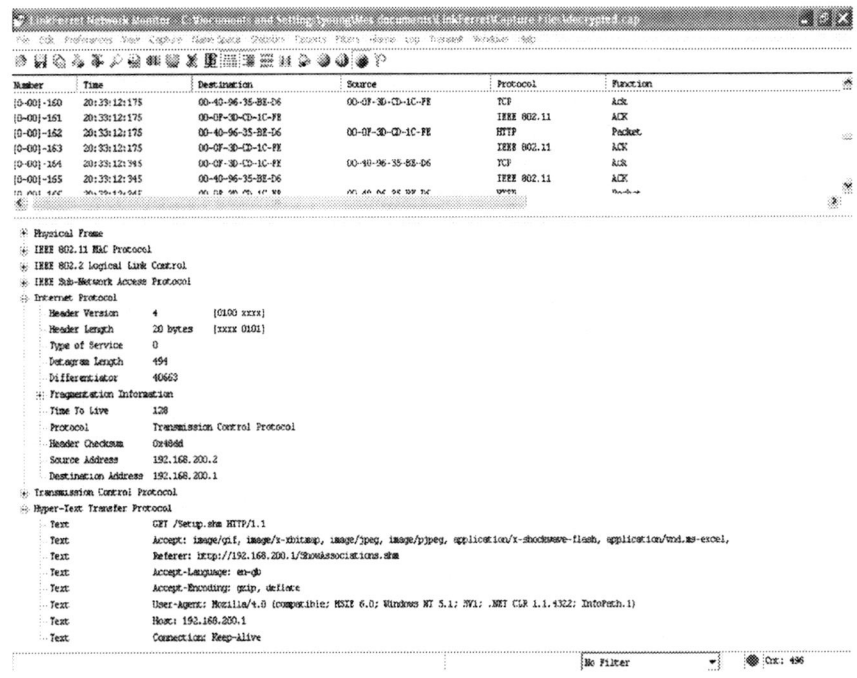

Fig. 6.5(a). LinkFerret screen shot showing captured frames and decoded HTTP messages

Data portion of the above packet can be decoded. In fact, as all subsequent packets are decrypted (still without the knowledge of the real client), many interesting observations are made. Suddenly the screen is transformed from a mass of meaningless data, to coherent HTTP packet exchanges, revealing the websites being browsed by the person being monitored. This view is shown in Figure 6.5(a).

Some of the data extracted from one decrypted packet is shown below. Here, the encrypted data is found to contain upper-layer protocol headers (for TCP/IP) and an HTTP request:

Internet Protocol
    Header Version   4   [0100 xxxx]
    Header Length   20 bytes   [xxxx 0101]
    Type of Service   0
    Datagram Length   494
    Differentiator   40663
    Fragmentation Information
    Time To Live   128
    Protocol   Transmission Control Protocol
    Header Checksum   0x48dd

Source Address	**192.168.200.2**
Destination Address	**192.168.200.1**

Transmission Control Protocol

Source Port	3645
Destination Port	80 Hyper-Text Transfer Protocol
Sequence Number	1843802233
Ack Number	2253027150
Header Size	20
Flags	[xx01 1000]
Window Size	17520
Checksum	0xc110
Urgent Pointer	0

Hyper-Text Transfer Protocol

Text	**GET /Setup.shm** HTTP/1.1
Text	Accept: image/gif, image/x-xbitmap, image/jpeg, image/pjpeg, application/x-shockwave-flash
Text	Accept-Language: en-gb
Text	Accept-Encoding: gzip, deflate
Text	User-Agent: Mozilla/4.0
Text	**Host: 192.168.200.1**
Text	Connection: Keep-Alive

Not only have the IP addresses of the client and the access point been discovered, the HTTP GET message is also intercepted. The page/Setup.shm is being requested from the host 192.168.200.1. Since this is the address of the AP itself, it is clear that the client is accessing the set-up pages for the wireless network.

Subsequent replies contain HTTP 200 OK messages. Since no upper-layer encryption is in place, the content of the web page is clearly transmitted. Due to the nature of the client's activities, in this scenario the entire configuration of the access point could be compromised.

Use of SSIDs. As it was previously described in paragraph 6.1.1.1, the concept of closed network is defined based on the fact that only users who know the name of the network or SSID are authorized to join the network. However, control frames containing SSID are sent in clear. Moreover, an access point can advertise its SSID, then a user can intercept it and access to the BSS. Even this advertisement option is disabled, used passwords are often simple and a dictionary attack allows to discover a great number of them. So, it is necessary to change the SSIDs.

Using the *LinkFerret Network Monitor* software it is possible to passively observe the packets present in any particular channel of the wireless medium.

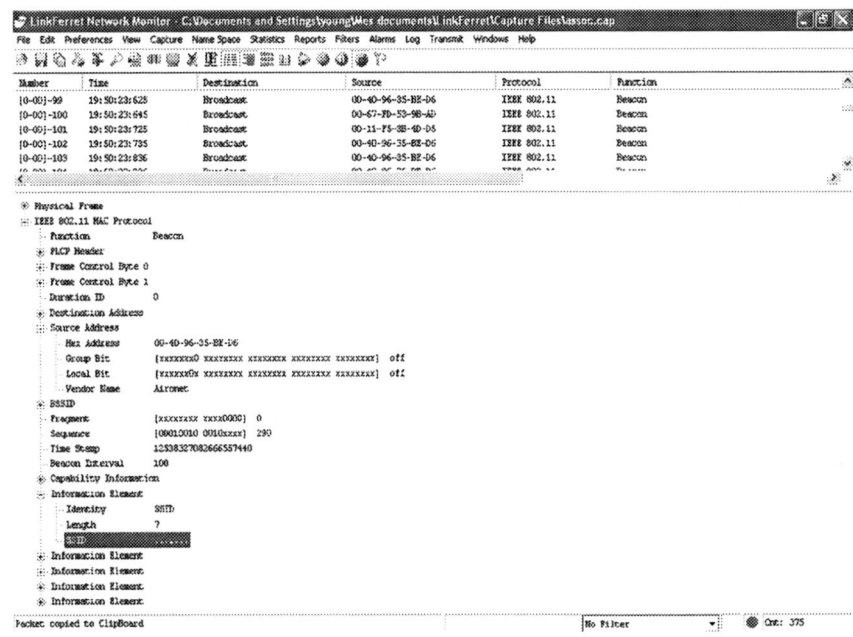

Fig. 6.5(b). LinkFerret screen shot showing captured beacon frames and masked SSID

A screenshot can be found in Figure 6.5(b), showing how this package can be used to capture packets and examine their contents, including the IEEE 802.11 protocol headers.

The beacon frames are broadcast periodically by all access points. An extract from one such frame illustrates that after attempting to cloak the AP, the SSID is indeed not included in these packets – although its length is evident:

> Information Element
> > Identity   SSID
> > Length   7
> > SSID   ....

As a result, the network no longer appears on the list of those available in Windows XP. The AP is even invisible to *NetStumbler*, software designed to discover all networks within the range of the wireless receiver.

However, the sense of security is soon discovered to be very false. As soon as a legitimate client wishes to connect to the network, a *probe request* is broadcast. Passive monitoring of the channel could miss this request if the client was out of range; however, being in range of the AP allows the network's response to be captured. This *probe response* frame is shown below, as interpreted by the *Ethereal Protocol Analyser*:

No.   82
Time   4.997186
Source   Aironet_35:be:d6
Destination   D-Link_cd:1c:fe
Protocol Info   IEEE 802.11 Probe Response, SN=285,
FN=0, BI=100, SSID: "tsunami", Name: "AP340-35bed6"

IEEE 802.11
   Type/Subtype: Probe Response (5)
   Frame Control: 0x0050 (Normal)
   Duration: 314
   Destination address: D-Link_cd:1c:fe (**00:0f:3d:cd:1c:fe**)
   Source address: Aironet_35:be:d6 (**00:40:96:35:be:d6**)
   BSS Id: Aironet_35:be:d6 (00:40:96:35:be:d6)
   Fragment number: 0
   Sequence number: 285
IEEE 802.11 wireless LAN management frame
   Fixed parameters (12 bytes)
   Tagged parameters (50 bytes)
      SSID parameter set: "tsunami"
         Tag Number: 0 (SSID parameter set)
         Tag length: 7
         Tag interpretation: tsunami
      Supported Rates: 1,0(B) 2,0(B) 5,5(B) 11,0(B)
      DS Parameter set: Current Channel: 12
      Cisco Unknown 1 + Device Name

The SSID can clearly be seen, transmitted in plaintext, along with other useful information about the network, such as the wireless channel used, and the supported data rates. Not only have the attempts to mask the network's identity been proved rather futile, but the ease of connecting to the network by legitimate users has also been seriously reduced.

ACLs. Another suggestion to control admission to a WLAN is to limit access by MAC address. These addresses are unique to any one particular Network Interface Card (NIC), and therefore tied to the identity of a unique, trusted user. However, whilst appearing to be fairly secure, MAC address filtering turns out to be completely superficial.

Access control based on MAC lists does not provide security for the two following reasons:

   – MAC addresses can be easily spoofed by an attacker because they are sent in clear, even if WEP is activated.

- The majority of the IEEE 802.11 wireless cards allow the change of their MAC addresses. Although engrained in the hardware of the device itself, the MAC address can be easily overridden by the operating system. This can be achieved using simple commands in Linux, by altering a registry entry in Windows, or even by using automated tools.

Consequently, an attacker can easily determine a list of the authorized addresses. It is thus necessary to maintain coherence in the presence of several access points.

Non-authenticated Control Messages. The following messages are not authenticated:

- Reauthentication
- Deauthentication
- Association
- Disassociation
- Beacons and probes

Note that TGW (802.11W) is working on improving the IEEE 802.11 MAC layer to increase the security of management frames.

1.3 Taxonomy of Attacks

Following the detected security faults, we establish a classification which gathers a large majority of the possible attacks.

Passive Attack by Traffic Spoofing. This attack is a direct consequence of the collision of the IVs. By exploiting the collision of two IVs, the result of XOR operation can be used to deduce the contents from the messages. In order to reduce the number of messages to intercept, the attacker can benefit from the IP traffic characteristics, which is generally foreseeable and includes many redundancies.

An extension of this attack would be to use a station connected to Internet and to send a traffic towards an IEEE 802.11 station. Knowing the original message and the encrypted message, the attacker will be able to recover the encryption key. Below, the following steps:

- One knows the coded message *ciphertext* and the message in clear *plaintext*.
- It was seen previously that *ciphertext* = *plaintext* XOR RC4 (IV,k).

- One can do *plaintext* XOR *ciphertext* = *plaintext* XOR *plaintext* XOR RC4 (IV,k).

- Like *plaintext* XOR *plaintext* = 0 (property of the XOR), we obtain RC4 (IV,k) = *plaintext* XOR *ciphertext*.

As long as the IV does not change, one is able then to decipher the messages by making *ciphertext* XOR RC4 (IV,k) = *plaintext*. It is then necessary to separate the message part of the CRC32 (*plaintext*).

Active Attack by Traffic Injection. This attack allows intercepting packets in transit, to modify them and to send them to the recipient without detecting that they were modified. The attacker must initially make assumptions on headers of the transmitted packets and not concerning the contents of these packets. If it manages to guess the IP address of the recipient, the attacker can change this address in order to direct the packet towards a machine which it controls on Internet. If the packet is correctly received by the access point, then it will be deciphered and directed, in clear, towards the desired machine. If, moreover, the attacker can guess the TCP header, then he will be able to change the destination port number of the packet (e.g. port 80) so that the traffic crosses the majority of the firewalls.

A second attack due to the linearity of CRC-32 is possible. If it is supposed that an attacker has a ciphertext and associated plaintext, it can then modify the ciphertext, calculate the CRC-32 and correct the integrity field ICV so that the packet is accepted. This attack is also possible even if the attacker does not know the whole plaintext. It can then carry out a selective modification of the data field and modify thereafter the CRC-32 value.

In this case, the attacker can:

- Obtain RC4 (IV,k) according to the preceding method

- Constitute its new message *newtext*

- Calculate the CRC32 (*newtext*)

- Concatenate two parts in *newplaintext*

- To send *newciphertext* = *newplaintext* XOR RC4 (IV,k)

It is important to note that the length of the secret key k has no impact on these attacks.

Dictionary Attack. The limited space of IV values as well as not very frequent change of the secret key gives the possibility to an attacker to built a table of keystream RC4 (IV,k) for each IV value. Once it learns the plaintext part of a few packets, it can determine the key generated by RC4. This key can be used to decipher all the packets using the same IV. This table requires up to 24 GB.

IP Redirection. The attacker seeks to guess the contents of the headers of packets in particular destination IP address. It can thus redirect the traffic, intended for this user, towards a machine under its control and which belongs to an Internet network. The packet will be successfully decrypted by the access point and transmitted not encrypted through the gateways and the routers until reaching the machine of the attacker.

Authentication without Keys. This attack is relatively simple. It is enough to follow the sequence of the following actions:

– Save a couple challenge/response with a sniffer

– Use the challenge to decrypt the answer and to find the key

– To use the obtained key to cipher the challenges

The attacker starts by intercepting the second and the third message of control during the exchange of message which takes place at the time of the authentication operation. The second message contains a random challenge sent in clear. The third message contains the challenge encrypted with the shared authentication key. The attacker thus knows:

– The random challenge P

– The challenge encrypted C

– The IV

The pseudo-random keystream is obtained by using WEP with the shared key k according to the following equation:

$$WEP^{(k,IV)} = C + P$$

The size of the pseudo-random chain represents the size of the authentication frame because all the elements constituting the frame are known: algorithm, number of sequence, code, length and text of the challenge. Morcover, the text of the challenge remains the same for all the authentication responses.

The attacker thus has the necessary elements to be authenticated with the target network without having to know the shared secret key k. It is enough for him to undertake the following actions:

– The attacker sends a authentication request to the wished access point

– The access point answers by sending a challenge in clear

- The attacker extracts the text from the challenge, pseudo-random stream $WEP^{(k,IV)}$ and forms the response text by carrying out an XOR operation of the two values
- The attacker calculates a new ICV value
- The attacker answers with a response authentication message, associates with the access point and joins the network

Denial of Service. This attack is based on the fact that the reassociation/disassociation messages are not authenticated. The mechanism of association allows association of a station with a particular access point. This mechanism is held after the authentication of the station. If this one wishes to change access point (reassociation) or to finish its association with the access point (disassociation), it is not obliged to authenticate itself. This point was exploited by the attackers who send a huge number of such messages to saturate the access and thus causing the denial of service (DoS).

Attacks Tools. By exploiting the weaknesses of RC4, tools such as WEPCrack, Airsnort and Sniffer Wireless were developed to crack WEP secret keys. These tools are downloadable on Internet starting from the following addresses:

- http://airsnort.sourcefourge.Net
- http://sourceforge.Netprojects/wepcrack
- http://www.sniffer.COMproducts/sniffer-wireless/

Obviously, there is a multitude of other means which make it possible to establish attack scenarios (ethereal, THC RUT, Kismet, LinkFerret, Netstumbler, AiroPeek, Airmagnet, Sniffer Wireless, Grasshopper, Mobilemanager, Airjack, etc.).

1.4 Diverse Solutions

All the completed work in the field of security in IEEE 802.11 networks highlighted a huge number of serious insecurity problems. This insecurity is due partly to the bad techniques of access control, but it results primarily from the WEP problems:

- Encryption with 40–128 bits keys
- Linearity of CRC-32
- Absence of the integrity control by message integrity check (MIC) with key

- Limitation of IV's space

- Absence of a protection against IV repetition

- Bad implementation of RC4 in WEP

- Bad management of keys

- Use of static keys (although manufacturers implement already the dynamic keys) even a single key for all the network was used in the beginning

Regarding the total failure of IEEE 802.11 security mechanisms, searching for immediate solutions was necessary. To mitigate these insufficiencies, two working groups were formed within the IEEE. Firstly, IEEE 802.11i designed a new protocol for data encryption and management of keys, while the second created the first grinding of the IEEE 802.1X, intended to ensure the security of the access to the network. While waiting for the standards, the manufacturers put on the market their first Wi-Fi products with some different implementations versions.

A panoply of solutions has been proposed which rapidly reach maturity giving the necessary methods to secure Wi-Fi/802.11x networks.

Following the same method adopted in paragraph 6.1.1, we will describe in this part, in a brief way, the various solutions with making an accent on the IEEE 802.11i standard which is the most important security driver of these networks.

Evolution of WEP to WEP2. WEP2 was proposed by IEEE 802.11 aiming at:

- Alleviating the insecurity limitations caused by the various WEP vulnerabilities

- Ensuring a compatibility with WEPv1

Mechanisms. WEP2 proposes the following functionalities:

- Increase the size of IVs to 128 bits (24 bits in WEPv1)

- Increase the size of the keys to 128 bits

- Periodic change of keys by using the IEEE 802.1X protocol

- Non-authentication of reassociate and disassociate messages

- Absence of protection against IV collision

- Use of RC4 algorithm

WEP2 Still has Flaws!. An analysis of WEP2 brings us to formulate the following conclusions:

- Since IV is broader in WEP2 (128 bits), the chances of IV collision decrease, but this does not completely prevent against intentional reuse.

- Attacks of decrypting traffic based on total or partial knowledge of the messages sent in clear, were not affected by the increase of IV size. Indeed, in the absence of a targeted protection against the reuse of IVs, it remains possible to recover data encrypted with the same key and to apply the attacks quoted in the preceding section.

- Authentication attacks remain possible because they were not affected by the IV size increase.

- Dictionary attacks are always possible.

- Attacks of decryption in real time become more difficult with WEP2 thanks to the increase of IV size, the memory necessary for the storage of a IV table would be of 2128 *1500 bytes = 5.11032 GB, not supported by the current systems.

- DoS attack remains possible since reassociate and disassociate messages are still not authenticated.

It is clear that WEPv2 does not solve all the problems of security posed by WEP and thus does not constitute an enough robust mechanism to constitute a complete security solution for Wi-Fi networks. Robust Security Network (RSN) the project supported by IEEE 802.11i aimed at enhancing such mechanisms. This project is based on the IEEE 802.1X protocol to provide a powerful solution to the problems of access control, authentication and key management.

Indeed, the IEEE 802.11i standard includes two classes of algorithms: pre-RSN security algorithms (WEP and authentication) and RSN algorithms.

MAC Ciphering Evolution and RSN Project under IEEE 802.11i standard.

The long-awaited IEEE 802.11i specification was fully ratified in 2004 and defines a new type of network – the RSN. The standard includes the following elements:

- Adoption of IEEE 802.1X authentication architecture (already present in Windows XP).

- Use of protocols Remote Authentication Dial-In User Service (RADIUS) or Kerberos for strong authentication and key management.

- Use of WEP, WEP2 with Temporal Key Integrity Protocol (TKIP), Counter with Cipher Block Chaining Message Authentication Codes

Protocol (CCMP) based on the use of Advanced Encryption Standard (AES).

- TKIP, short-term solution intended to improve the mechanism of authentication per packet and to replace ICV mechanism.
- AES, this new ciphering standard will replace, in the long term, completely the encryption algorithm to reach a very good level of security and to carry out a more powerful integrity control.
- Use of Michael as a control integrity code MIC, on 64 bits, to ensure the integrity of the messages.
- Definition of a hierarchy for the generation of keys.
- Use of a packet sequence number.

TKIP can be directly supported on the existing equipments by changing only the software. Contrary to TKIP, AES is incompatible with the existing products.

Another aspect is devoted to fast mobility (fast handoff) secured mechanisms where reauthentication is not needed, security profiles between various access points during roaming are defined as it is defined in IAPP.

Temporal Key Integrity Protocol. Protocol TKIP, using WEP, integrates three new aspects:

- Use MIC instead of CRC.
- New ordering of IVs: a security flaw that MIC cannot detect is the replay attack. The best way of circumventing this problem is to assign sequence number to each packet. TKIP uses IV as a packet sequence number. Transmitter and receiver initialize this number to zero at each generation of new keys. The transmitter increments this number with each packet sent. At reception, a packet is considered valid if its sequence number is strictly higher than the number of a correctly received packet associated to the same key.
- Cryptographic function allows to combine for each packet the temporal session key, the sequence number and the IV. This mixing function thus permits to obtain a 128-bit key per packet. This key per packet is necessary to avoid the reuse of the keys which was a major problem in WEP. The temporal session key is derived starting from the key (master key) which is used in IEEE 802.1X architecture.

Procedures of encryption/decryption are described respectively in Figures 6.6 and 6.7.

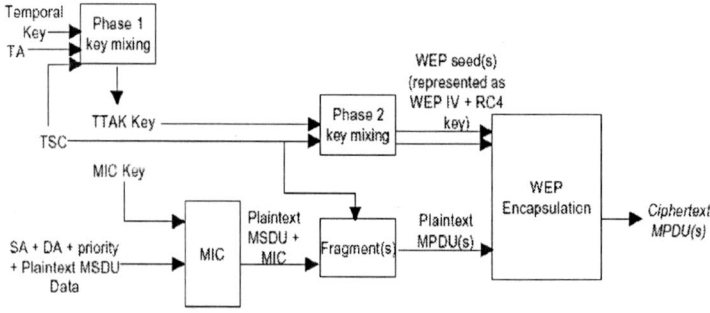

Fig. 6.6. Encryption/encapsulation with TKIP

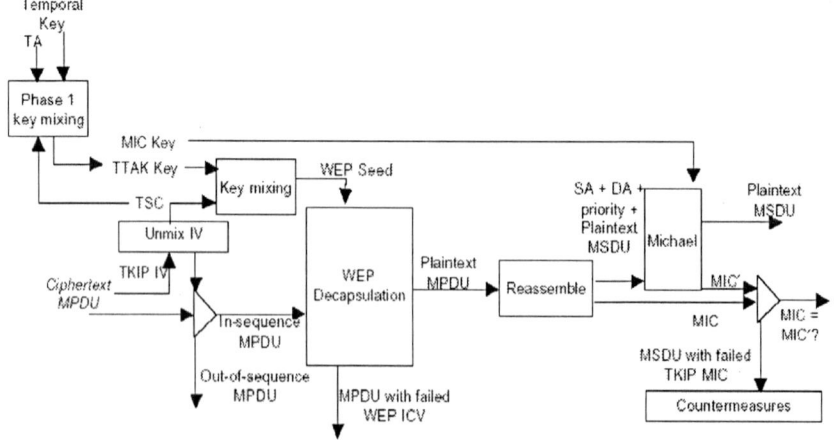

Fig. 6.7. Decryption/decapsulation with TKIP

TKIP improves WEP by adding some functions, as shown in Figure 6.8:

– TKIP calculates MIC (called Michael) of a MSDU of 64 bits length, including the following elements: source address, destination address, priority and data field. It concatenates MIC with MSDU.

– TKIP splits up the MSDU in several MAC Protocol Data Units (MPDUs). A sequence number TKIP sequence counter (TSC) is assigned to each MPDU. It is the 48-bit extended IV. TSC allows to lengthen the lifespan of the session key eliminating the need to recompute the key lasting a security association (SA).

– For each generated MPDU, TKIP uses a mixing cryptographic function to generate a key. This function, operating in two phases, uses parameters

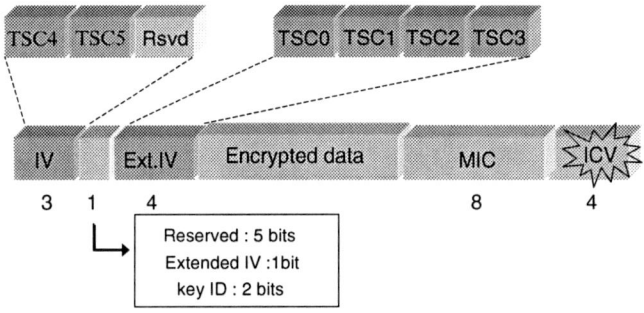

Fig. 6.8. Encrypted frame structure with TKIP

like TSC, the temporal key (shared key secret on 128 bits) and MAC address of transmitter MT (transmitter address).

- WEP encryption of each MPDU is carried out by using RC4 key (104 bits) and IV (24 bits). A WEP key generated per packet is thus associated to a temporal session key. This key and MIC key are derived during the IEEE 802.1X exchanges.

TKIP can have the following countermeasures:

- Exhaustion of the sequence number TSC. If a new temporal session key cannot be defined before the exhaustion of the space of the values of sequence numbers (on 16 bits), then communications will have to stop.

- Failure of keys refresh. The implementation envisages to stop considering the traffic until the process of keys regeneration finishes successfully or a disassociation will take place.

- Failure with MIC. The station erases the keys and reassociates.

WEP decryption procedure is improved thanks to the execution of the following operations:

- Before decapsulating a received MPDU, TKIP extracts the sequence number TSC and the key from IV. If the sequence number is valid, TKIP uses the mixing function to derive the WEP key otherwise the packet is rejected.

- TKIP sends (IV, RC4 key) and MPDU fragment to the decapsulation module.

- If ICV value check by WEP proceeds successfully, MPDUs are thus reassembled in a MSDU. A MIC integrity control will take place thereafter.

– If the integrity check of the whole MSDU (source address, destination address, priority, data field) fails the packet is rejected. In the contrary case, the received MSDU is delivered. Countermeasures, quoted above, are called upon if the frame is erroneous and to alert if necessary the network administrator.

Because of MIC weaknesses, TKIP takes the following actions. If the integrity control fails, the used keys for encryption and integrity will not be used any more. The failure rate of MIC procedure should be minimal. Aiming at decreasing the risk of false alarms, the integrity control is carried out after the other types of controls (CRC, IV).

The structure of TKIP MPDU packet is illustrated by Figure 6.8. TKIP extends the encrypted WEP frame of 12 bytes, 4 bytes for the extended field IV and 8 bytes for the integrity code MIC.

Because of the increased size of MPDU frames (because of the TSC and field MIC), MSDU (MAC source address + MAC destination address + data) of maximum size of 2312 bytes can be fragmented in MPDU of 2346 bytes maximum size. The maximum sizes actually depend on the adjustment of Fragment_Threshold parameter, by default set to 2346 bytes.

TKIP is an intermediate Solution. It is not viewed as a long-term solution, like AES, nevertheless it remains better than WEP. In Sections 6.1.4.4 and 6.1.4.5 we give a description of AES and CCMP.

AES. AES was developed by National Standard and Technology Institutes (NIST) in 1997. It is an iterative per-block ciphering algorithm which supports various combinations [length of key] – [block length]: 128-128, 192-128 and 256-128 bits.

AES operates on blocks of 128 bits (plaintext) transformed into encrypted blocks of 128 bits (ciphertext) while following a sequence of Nr operations or "rounds", starting from a key of 128, 192 or 256 bits. According to the size of this latter, the iteration count differs; it can be 10, 12 or 14.

A brief description of encryption/decryption AES is available at http://en.wikipedia.org/wiki/Advanced_Encryption_Standard.

Among the strong features of AES, we find simplicity/speed of calculation, weak resources, implementation flexibility (software or hardware), resistance to linear cryptanalysis attacks and robustness to dictionary attacks (Figure 6.9).

CCMP. Counter-Mode CBC-MAC Protocol (CCMP) is based on AES in mode Counter Mode with CBC-MAC (CCM). This last combines CTR mode (Counter) ensuring confidentiality and authentication/integrity CBC-MAC [HOU 03] (Cipher block Chaining Message Authentication Codes). Only one key is used for the two procedures. Like TKIP, CCMP uses an extended 48-bit

Security in WLAN, WPAN, WSN and WMAN

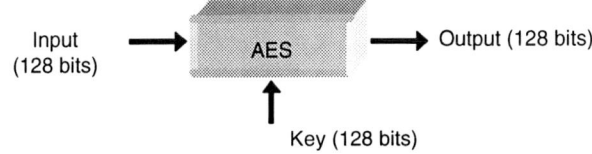

Fig. 6.9. An AES block (used by CCMP)

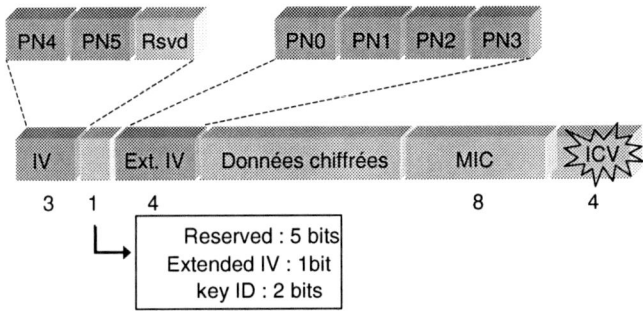

Fig. 6.10. CCMP encrypted frame structure

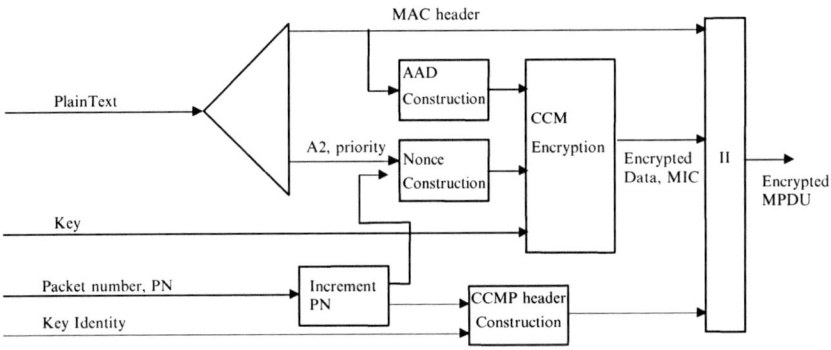

Fig. 6.11. Encryption/encapsulation with CCMP

IV called Packet Number (PN), instead of TSC, whose position in the frame remains unchanged.

ICV field is owed but the encrypted frame MPDU format remains identical to the TKIP frame; it is shown by Figure 6.10.

The process of encapsulation is illustrated by Figure 6.11. CCM, packet-oriented, is based on a block ciphering. Encrypted AES blocks defined as well for MIC calculation as for data ciphering use the same temporal session key (noted K). Similarly to TKIP, the temporal key is derived from one key (master key) generated by IEEE 802.1X. Ciphering and MIC operations are launched in parallel. Several elements contribute to calculate MIC (PN, extracted header information, useful data). A first AES block receives as input PN, the key, whose output is concatenated with some elements of the header (additional authentication data [AAD]) and a nonce; which is transmitted like input data to another AES block. The nonce is calculated from some information (packet number, destination address, priority). This process continues until processing all data of the frame (header included).

The encryption algorithm CTR (Counter) encrypts data and the obtained MIC; it uses a sequence number determined from some elements (PN, a flag, data extracted from header, counter set to 1). This sequence number is used (to calculate the first AES block which output is XORed with a 128-bit block of a non-encrypted frame.

The counter is incremented and the process is thus repeated for the following stored block of 128 bits until treating the totality of the useful data. The final counter is reset to 0 and injected as input to an AES block which its ouput is XORed with the MIC calculated previously. Encrypted MIC field is then concatenated with the encrypted frame to constitute the complete frame which will be finally transmitted. This process is described by Figure 6.12.

Decryption/decapsulation procedure takes place in an opposite way to that of encryption/encapsulation. A last step is added in order to compare MIC and AAD values to those received before the decrypted frame has been delivered to the MAC layer.

Table 6.1 summarizes the essential characteristics of the three suggested protocols.

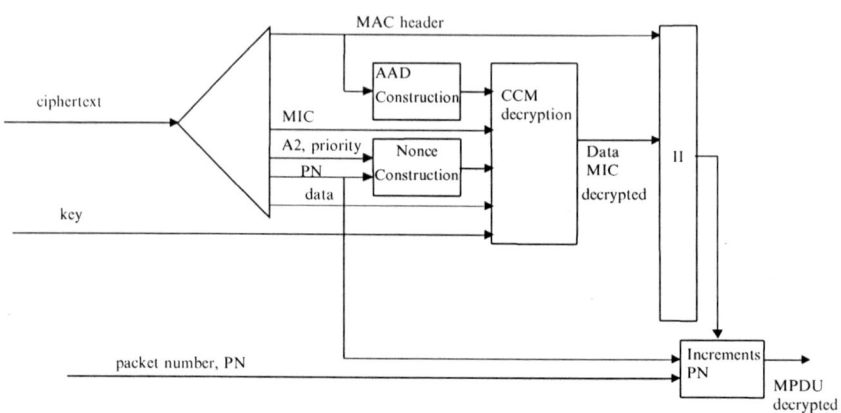

Fig. 6.12. Decryption/decapsulation with CCMP

Table 6.1. Comparative table between WEP, TKIP and CCMP

Protocol	WEP	TKIP	CCMP/AES
Encryption	RC4	RC4	AES-CCMP
Size of key (bits)	40 or 104	128 encryption 64 authentication	128
Key liveness	IV 24 bits	IV 48 bits	IV 48 bits
Key per packet	concatenation	Mixing function	Not necessary
Data integrity	CRC32	Michael	CCM
Header integrity	No	Michael	CCM
Replay	No	Use IV	Use IV
Key management	No	IEEE 802.1X	IEEE 802.1X

Also, IEEE 802.11i indicates three building blocks to carry out SA management. It concerns:

- The use of RSN negotiation to establish a security context. For example, a BSS can support several encryption protocols the access point and the user must pass by a negotiation phase to choose the highest level of security being able to be supported by both.

- Authentication based on IEEE 802.1X and encryption (TKIP or AES).

- Key management by means of IEEE 802.1X providing the various keys.

802.1X. IEEE 802.1X provides a framework for strong mutual authentication and key distribution. IEEE 802.11i defines the interactions between IEEE 802.1X and 802.11.

IEEE 802.1X authentication uses three entities:

- A customer (Supplicant) wishing to access to a service.

- An authenticator (which generally corresponds to the access point). It relays authentication frames between the client and the waiter of authentication. From the point of view of the client, the authenticator carries out the authentication.

- An authentication server: this server communicates using an Authentication Authorization Accounting (AAA) protocol with the authenticator. RADIUS [RIG 00, WIL 00] servers are the most used.

Figure 6.13 shows the various entities implied in 802.1X.

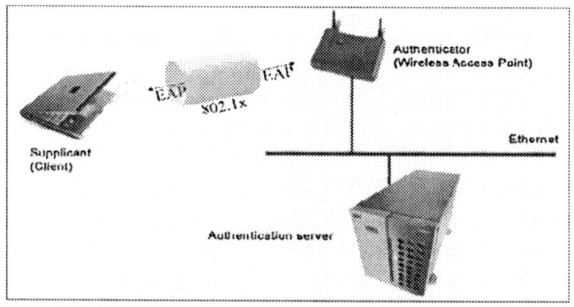

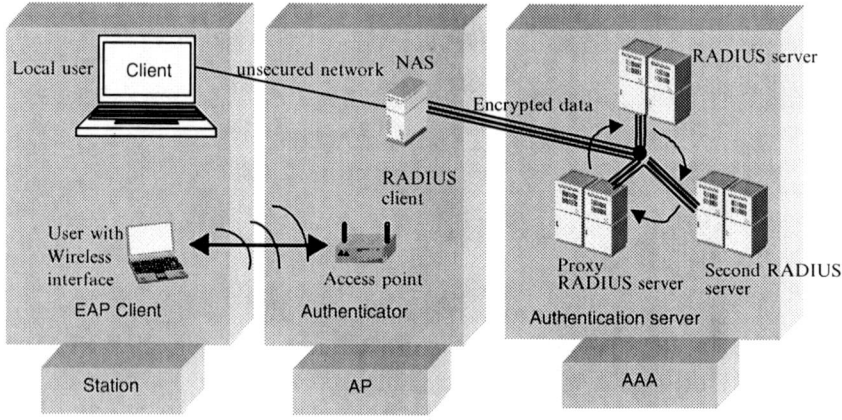

Fig. 6.13. IEEE 802.1 IX architecture

IEEE 802.1X requires the cooperation of two protocols:

- An authentication method, negotiated on level 2, thanks to the Extensible Authentication Protocol (EAP) [BLU 98], between the client and the authenticator. Several protocols are candidates such as EAP-Message Digest 5 (MD5), EAP-Transport Layer Security (TLS) [ABO 99], lightweight EAP (LEAP) and tunnelled TLS (TTLS). EAP-TLS automatically generates encryption keys which can be renewed periodically and makes them thus appropriate to be used within IEEE 802.11.

- An application-level AAA protocol, used between the authenticator and the authentication server, in order to allow the authentication of the user, its authorization with the access to the services and the billing of its services.

An IEEE 802.1X dialogue is performed according to the diagram of Figure 6.14.

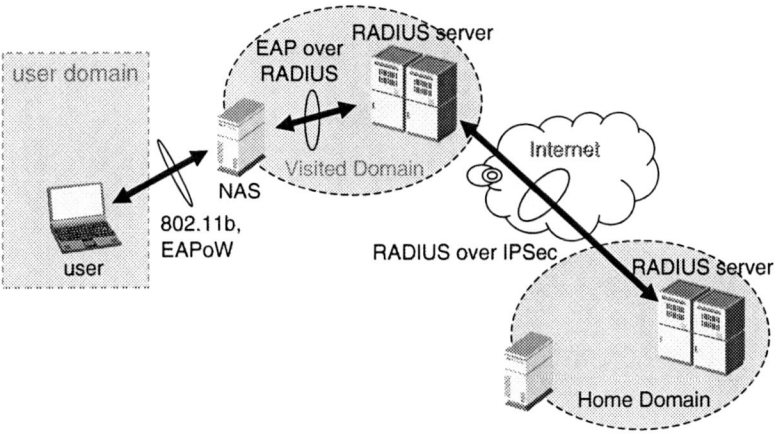

Fig. 6.14. IEEE 802.11 and IEEE 802.1X

The authenticator sends an EAP-Request/Identity packet to the client as soon as it detects that the link is in activity, it asks him to be identified. Supplicant sends an EAP-Response/Identity packet to the authenticator which transmits it to the authentication server. The authentication server answers with a challenge sent to the authenticator. The authenticator encapsulates the EAP packet in an IEEE 802 frame and sends it to the client. Various authentication methods can be used. EAP supports two types of authentication: simple and mutual. The mutual method is suitable in the case of wireless networks. In this case, the client answers the challenge via the authenticator and passes the response to the authentication server. If the client delivers the correct identity, the authentication server answers with an EAP-success message which is then sent towards the client. The authenticator then allows the client to access to the network according to the attributes set by the authentication server. If the authentication fails, the authentication server informs the authenticator that breaks the link with the client.

This dialogue requires the encapsulation of protocol EAP in IEEE 802 frames. We call EAP Over LAN (EAPOL) the protocol deployed in fixed LANs and EAP Over Wireless (EAPOW) in IEEE 802.11. This encapsulation requires some extensions of EAP in order to integrate IEEE 802 frames.

The authenticator can also transmit a message EAPOL-key or EAPOW-key to the client, indicating information on a key to be used to protect the started session.

The global architecture is shown in Figure 6.14.

A typical authentication using the protocol 802.1X in a network IEEE 802.11 is illustrated by Figure 6.15a.

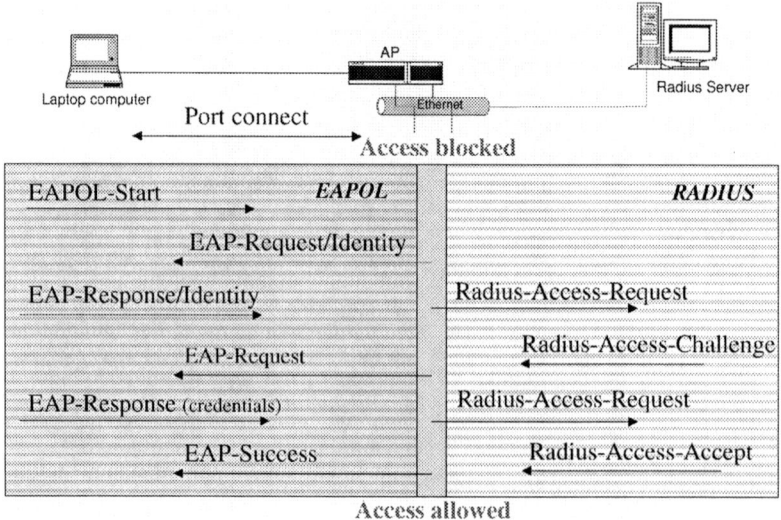

Fig. 6.15a. Authentication by IEEE 802.1X in an IEEE 802.11 network

Especially, choosing EAP-TLS within IEEE 802.1X framework permits to considerably enhance Wi-Fi networks security (see Figure 6.15b). Indeed EAP-TLS is an open standard supported by most of the existing wireless equipments. Based on EAP, it uses the concept of public key infrastructure (PKI). A user must have a certificate to be able to authenticate itself with network. The AAA server must also have a certificate to validate its identity regarding its customers. A certification authority (CA) delivers certificates to both AAA servers and users.

PKI with EAP-TLS is a solid solution since private and public keys are used. The exchange of EAP-TLS messages offers a mutual authentication, a common negotiation of encryption methods, and a secure keys exchange between client and authentication authority. EAP-TLS offers for the moment the most secure method for both authentication and keys exchange.

We notice that IEEE 802.1X authentication in IEEE 802.11 systems, occurs after the phases of authentication and association. Moreover, the exchange of WEP keys is possible only if IEEE 802.1X authentication is successfully achieved.

Thanks to IEEE 802.1X, the following security mechanisms are provided:

- An authentication and an integrity of each packet exchanged between the RADIUS server and the access point. Indeed, these two entities share a secrecy. This secrecy is used to authenticate the messages exchanged between RADIUS server and the authenticator (thanks to the attribute Message-Authenticator) and between the authenticator and RADIUS

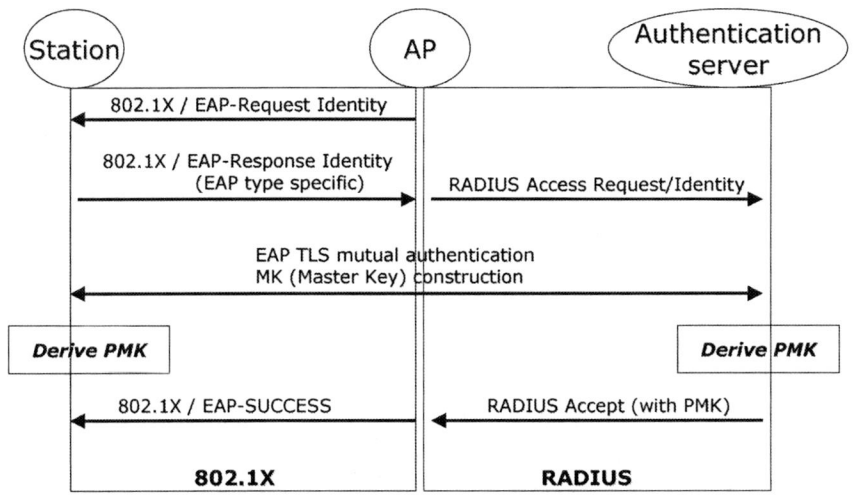

Fig. 6.15b. Authentication with EAP-TLS

server (thanks to the EAP-Authenticator attribute). Consequently, a mutual authentication between these two entities is carried out.

- Flexibility and scalability are provided. Indeed, by separating the authenticator from the authentication server (the authenticator does not have any more the responsibility of authenticating itself the client it just have to transfer authentication requests to the server), the number of access points can be increased easily.

- Flexibility based on confidentiality support (attribute EAPOL key message).

- Access control: an authentication based on user identity constitutes a rather effective access control. However, attacks such as session hijacking are possible.

- A flexibility of key management thanks to the possibility of changing frequently and dynamically keys by the access points or authentication server.

Remote Authentication Dial-In User Service. RADIUS is specified by the IETF and designed for use over TCP/IP networks. It defines a set of functionalities that should be common across authentication servers, and a protocol that allows other devices to access these capabilities.

There are four core RADIUS message types: *Access-Request, Access-Challenge, Access-Accept and Access-Reject.* All EAP messages are encapsulated in the *attribute* field of RADIUS. If the AP does not know if the RADIUS server supports EAP, it can send an *EAP-Start* message, within

Code	Identifier	Length
1 octet	1 octet	2 octets
Data		

Fig. 6.16. EAP packet format

an *Access-Request*. The response is either *EAP-Request/Identity* or a reject message.

In a wireless environment, it is important to authenticate every message within the authorization process. This prevents the identity of a valid user, having been granted access to the network, from being stolen (known as *session hijacking*). To avoid this, the server sends a master key to the AP to provide per-packet authorization and integrity protection. This key is itself encrypted and sent in a key delivery attribute.

Extensible authentication protocol. EAP is specified in RFC 2284 as part of PPP. Originally it was designed as an alternative to CHAP/PAP protocols. It is a very simple and extensible protocol. It transports various authentication protocol messages between a user (client) and an authenticator server. A number of EAP methods have been defined to address several requirements in WLAN security. We can find EAP-MD5, EAP-TLS (widely used), EAP-TTLS, PEAP, EAP-SIM, EAP-AKA and EAP-fast. These authentication methods comprise certificate and password-based authentication but also permit the usage of GSM SIM cards, SecureID tokens, etc. For example, EAP-MD5 is a weak authentication method since it only offers client-side authentication. In addition to MD5 weaknesses (collision attacks), passwords are stored in clear on the AAA server and some AAA traffic attacks are possible. EAP-TLS provides highest security support using digital certificates at both the client and the AAA server. LEAP, a Cisco solution offering an effective security solution using WEP-based equipments. But it is vulnerable to dictionary attacks.

EAP thus covers a vast variety of the popular user equipment. Since its introduction for IEEE 802, EAP has gained a tremendous popularity.

The structure of the EAP packet is given in Figure 6.16. It comprises five fields: code, identifier, length, type and data.

WPA and RSN Key Hierarchy. In this section we clarify the roles of the many different keys involved in the involved security architecture. Keys fit into one of two key hierarchies, known as the *Pairwise Key Hierarchy* and the *Group Key Hierarchy*. At the top of each is the master key, which can take one of two forms: *preshared* or *server-based*.

Preshared and Server-based Keys. A preshared key (PSK) is manually installed in the mobile device and the AP by some method outside of Wi-Fi protected access (WPA) or RSN. Note that this bypasses upper layer authentication, and identity is assumed by possession of the key. This is particularly suitable for deployments in the home or small office environments, where RADIUS implementations are not warranted or are deemed too expensive. This system is commonly marketed as *WPA-PSK* or *WPA-Personal*.

In WPA systems, the PSK is 256 bits in length, or 64 hexadecimal characters. However, it is difficult for a user to enter such a key directly. Instead, a pass-phrase can be used, composed of 8–63 ASCII characters. This phrase is hashed together with the SSID to derive the 256-bit PSK. However, care must be taken to choose an appropriately secure pass-phrase.

A more robust, expandable solution uses a server-based key. This employs an upper-layer distribution process, thus omitting the human element and allowing for the use of more secure keys. The key is generated by the mobile device and the authentication server, which passes it to the AP through a secure connection. This system is suited to large WLAN deployments, and as such has been marketed as *WPA-Enterprise*.

The two key hierarchies are almost identical for both TKIP and AES-CCMP encryption schemes, used in WPA and RSN, respectively. However, slightly fewer keys are required in the latter, for reasons that will become clear. In total, TKIP requires 768 bits of storage for its transient keys, whereas AES-CCMP reduces this to 512 bits. To ensure interoperability in an environment of mixed equipment, TKIP should be used for multicast. If AES-CCMP is used for multicast, TKIP cannot be used for pairwise security.

Pairwise Key Hierarchy. These keys protect communication between a single pair of devices, for example, a mobile device and an AP. The AP must store a pairwise key for each and every mobile device.

At the top of the hierarchy is a 256-bit (32-byte) pairwise master key (PMK). This can be either a PSK, or delivered from an upper-layer protocol. The PMK is used to derive the temporal keys, which together make up the pairwise transient key (PTK). These temporary keys are recomputed every time a device associates with an AP.

The PTK is generated from the PMK, combined with a nonce and an MAC address from both authenticator and supplicant. In this way the PTK is bound to the identity of both devices (Figure 6.17).

The PTK under TKIP consists of four keys, each 128 bits in length:

- *EAPOL-Key*: key conformation key (KCK)
- *EAPOL-Key*: key encryption key (KEK)

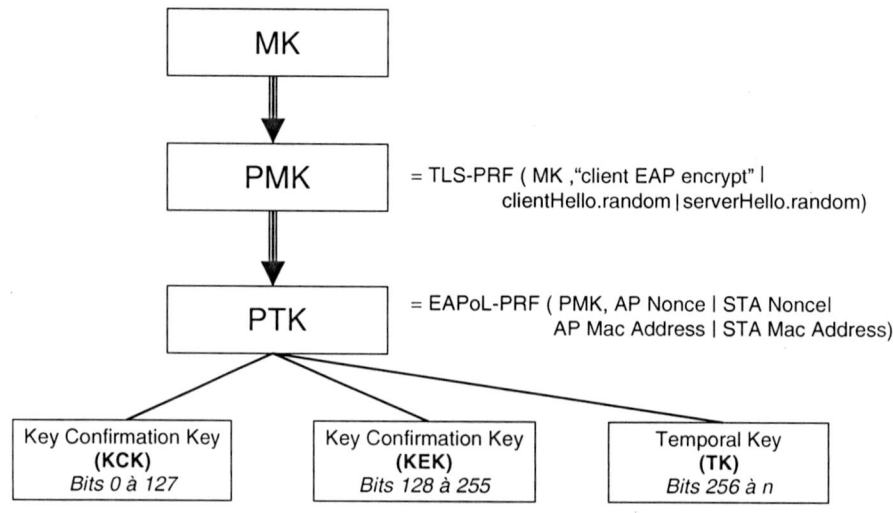

Fig. 6.17. Key hierarchy

- Data encryption key

- Data integrity key

The KCK is used ensure to check authenticity and integrity by calculating a MIC, during the four-way handshake and group key handshake procedures, described later. Similarly, the KCK is used to provide confidentiality of the *EAPOL-Key* frames during these exchanges.

In AES-CCMP, data integrity and encryption functions are combined, so do not require different keys. Therefore, the last two keys above are combined into one, 128 bits in length.

Four-way Handshake. The methods of distributing keys apply to both TKIP and AES-CCMP; the only difference is the key lengths as mentioned (Figure 6.18a). In general, the AP and the mobile device must prove they both know the PMK, so that they can be mutually trusted. To do this, a *four-way handshake* is exchanged between the Authenticator (A) and the Supplicant (S), using *EAPOL-Key* messages:

1. $A \rightarrow S$: Delivers A's MAC address and *ANonce*; unencrypted

    - This allows S to compute the PTK
    - The KCK can now be used to compute the MIC in subsequent message, to allow the detection of tampering

Security in WLAN, WPAN, WSN and WMAN   191

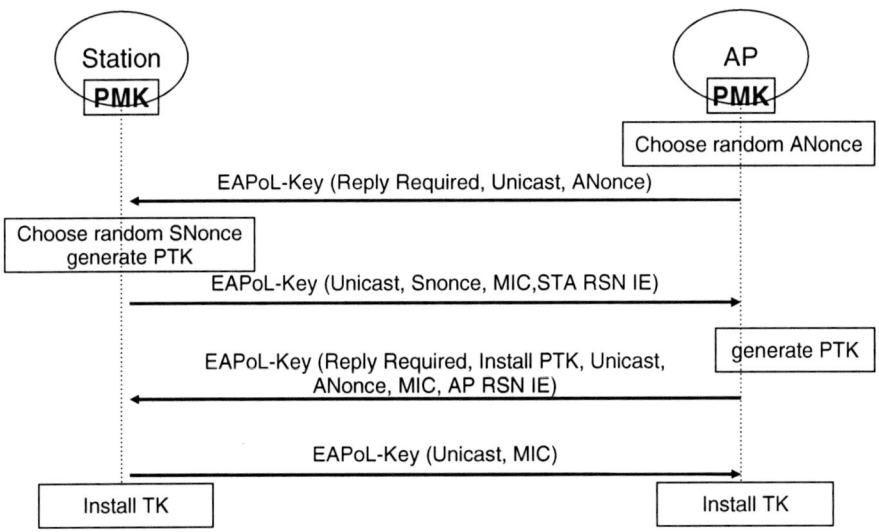

Fig. 6.18a. Four-way handshake

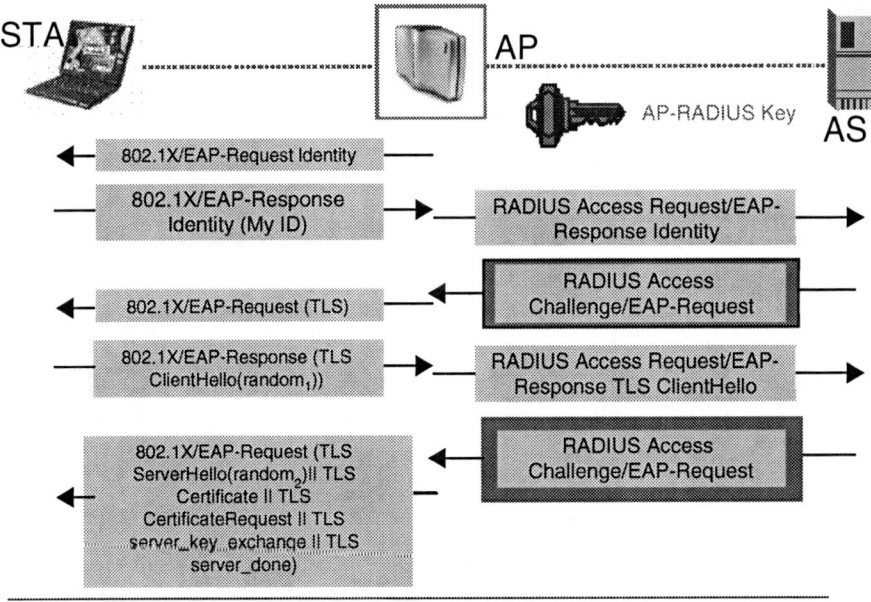

Fig. 6.18b. Four-way handshake with TLS EAP authentication

2. $S \rightarrow A$: Delivers *SNonce*; unencrypted, but tamperproof

- This allows A to compute the PTK
- A can then verify the MIC. This proves that S knows the PMK

3. $A \rightarrow S$: Signals A is ready to begin using new keys for encryption
   - Cannot be replayed, because MIC is computed from *SNonce*
   - Unencrypted, and proves that A knows the PMK
4. $S \rightarrow A$: Unencrypted acknowledgement
   - S installs keys after sending, A after receiving this message

Figure 6.18b shows the four-way handshake with EAP-TLS use. The disadvantage is that a four-way handshake can be slow, sometimes taking several seconds. This could present problems in some circumstances, such as executing a sufficiently fast handover between APs during VoIP calls.

Group key Hierarchy. A group key is shared by all members of a trusted group, and is used for multicast or broadcast messages (Figure 6.19. It must be changed whenever a member leaves the network, to ensure that previously networked devices can no longer receive broadcast messages.

The AP creates a 256-bit cryptographic-quality random number (i.e. it is not at all predictable) known as the group master key (GMK). This does not need to be bound to any identity, as authentication will not be required.

The AP then derives the group transient key (GTK). As was the case with the PTK, only one of these keys is required under AES-CCMP, 128 bits in length; in TKIP, both 128-bit keys are required:

- Group encryption key
- Group integrity key

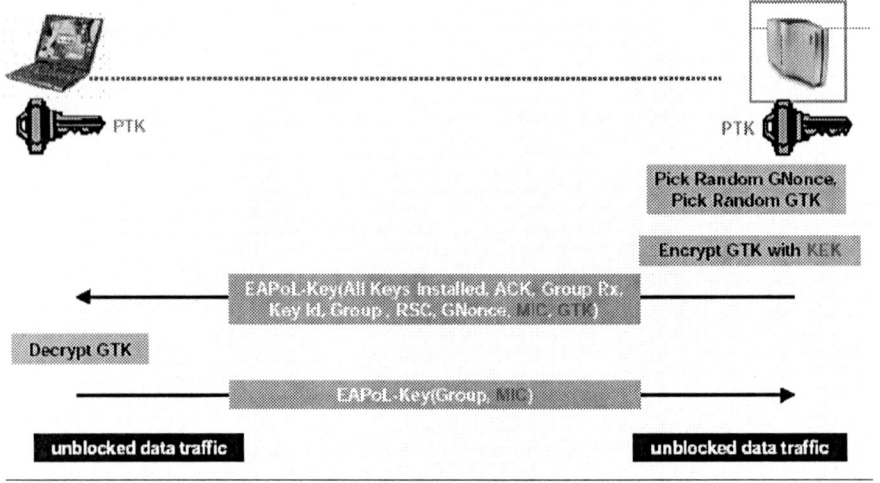

Fig. 6.19. Group key handshake

After establishing the pairwise secure connection, the AP can then send the GTK to each device. This is achieved using the *group key handshake*, performed through the following exchange of *EAPOL-Key* messgaes:

1. $A \rightarrow S$: Delivers the GTK

    - The key data itself is encrypted using the KEK
    - The KCK is used to calculate a MIC, protecting the entire message from tampering

2. $S \rightarrow A$: Unencrypted acknowledgement

    - The KCK is again used to calculate a MIC

Replay counters are also included in the messages to protect against replay attacks. To avoid service disruptions when changing keys, the devices in fact store two GTKs, one key currently being used, and one new key. The new one is only used when all the nodes in a network have acknowledged its receipt.

VPN, Firewall and Application-level Security. For wide area networks with strong security requirements, the solution of virtual private networks (VPN) is advised. Virtual tunnels are configured to ensure that only the authorized users are able to reach the corporate network. Several approaches are possible to implement VPNs where two aspects are important: authentication and data encryption. It is necessary to consider operations of reauthentication during roaming between VPNs. The most widespread protocols are Point-to-Point Tunneling Protocol (PPTP), Layer 2 Tunneling Protocol (L2TP) and IP security (IPsec) [LOS 00].

Firewalls can filter the packets per address and/or by port. In the case of wireless networks, a firewall is put behind each access point or in the equipment itself.

Securing applications is also possible. Several protocols can be used such as Secure Socket Layer (SSL), TLS, Secure Shell (HS) and Pretty Good Privacy (PGP).

Another solution can be the installation of intrusion detection systems with servers IIS, Real Scoure IDS or Air IDS. At access points, it is necessary to have SNMP traps and login-based mechanism.

1.5 WPA/WPA2: IEEE 802.1X and TKIP/AES

WPA is a label of Wi-Fi Alliance to certify the Wi-Fi equipments implementing a subset of IEEE 802.11i mechanisms such as IEEE 802.1X or TKIP. WPA was designed to work immediately with existing hardware. It was shipped with most IEEE 802.11g devices, or available as a firmware upgrade for older equipment.

WPA defines two operating modes; "SOHO" mode and "Enterprise" mode. The first requires little configuration and is adapted to small companies or domestic use which requires only little security while the second is more robust and implements more complex mechanisms. In response to the final ratification of the IEEE 802.11i specification, the Wi-Fi Alliance updated their WPA standard to WPA2, a full implementation of the RSN that usually requires a full hardware upgrade.

SOHO Mode. This mode is based, like WEP, on a secret key shared by all users or simply between the user and the access point. This key is used for authentication, encryption via TKIP mechanism and MIC-based integrity.

"Enterprise" Mode. This second operating mode also uses TKIP and MIC. But authentication is more complex by using an IEEE 802.1X architecture.

1.6 Synthesis

Richness and complexity of the topic of security in IEEE 802.11 WLANs, illustrated throughout this part, lead at the end of the exhaustive state of the art which we carried out, to enumerate a certain number of directives, some are short, other long-term, to set up a secured system.

Because of the failures of the first basic mechanisms, some attitudes appeared such as:

- Limit, as far as possible, the area radio range into a zone or into private buildings which are under control of the owner of the network, so as to avoid unauthorized access from a public zone.

- Configure each access point with a personalized SSID different from that defined by default. The mode broadcast SSID must be enabled. It is recommended to regularly change SSIDs.

- Filter MAC addresses.

- Activate WEP in spite of its failures, to use keys of 128 bits and change regularly WEP keys. To deriviate WEP keys from a main session key and to use dynamic keys.

- Monitor the sequencing of long IVs.

Aiming at reinforcing security, according to needs, we need to:

- Set up the IEEE 802.1X authentication architecture

- Filter packets via a firewall

- Use application security protocols (IPsec)

Security in IEEE 802.11 networks is a field which has evolved and reaches a maturation state. For medium-term, the standard IEEE 802.11i, with TKIP and/or AES will be integrated in the equipment. It is obvious that VPNs with IPsec tunnels represents a robust solution.

Since ad hoc networks and WMNs are more and more deployed, security issues related to this kind of configurations have to be investigated. We have to design accurate trust models and to establish appropriate trust management mechanisms.

2. Security in Bluetooth Systems

In addition to lower rates compared with those offered by Wi-Fi, Bluetooth suffers from another handicap which is security. Bluetooth security requirements depend on three factors:

- Strategic character of exchanged data: in the case of a domestic PAN connecting very close devices, data confidentiality is not essential. One may always acquire necessary material to divert data, but is this really an objective?

- Probability of attacks: data in a wireless network can be critical, since they impact the correct operation of equipments, even the security of home environment: fire protection alarms, alarms, water leakages detection, etc.; domestic networks have to be correctly secured.

- Services: typically, security needs is more important for professional uses than for personal uses. The Intranet of a company should be well protected.

Coming from telephony world, Bluetooth security was inspired by that existing in second-generation DECT digital phones. It is in fact a point-to-point security by exchange of private keys.

Taking into account the fact that Bluetooth uses a radio connection, the broadcasted traffic can potentially be captured by the neighbouring receivers. Even if Bluetooth uses fast FHSS, MAC authentication and authorization mechanisms are used. Bluetooth supports a series of means to offer security at both lower and higher layers.

We detail the specified procedures in the Specification 1.1, in order to analyse their operation and to show their vulnerabilities.

2.1 Security Architecture

The reference model, illustrated by Figure 6.20, includes a security dedicated entity called Security Manager [MUL 01].

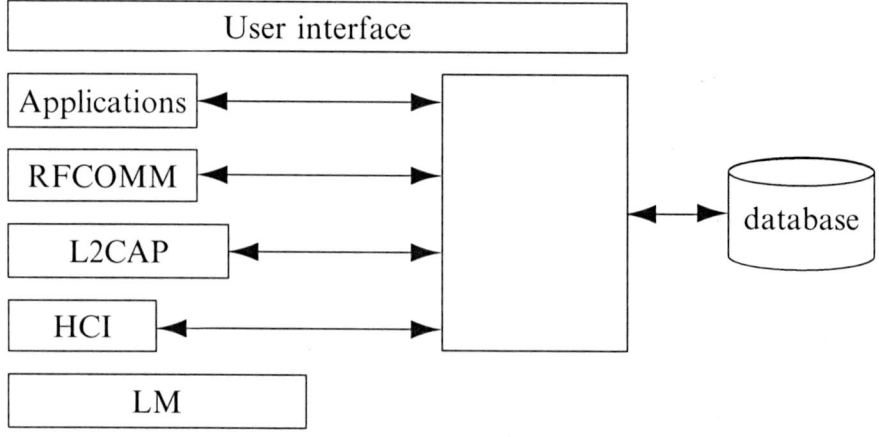

Fig. 6.20. Bluetooth reference model

Before the establishment of a connection, the security manager decides which security policy to apply. This decision is based on the used service type and the distant device with which the communication will take place (its type and its security level). Security information needed by the security manager is stored in two databases: the database of the physical devices (which stores the device type, its reliability level and size of its encryption keys) and another for services (information related to service security level and data routing mode).

Security Levels. Bluetooth defines various security levels for peripherals and services. Each peripheral obtains a status when it connects, for the first time, to another peripheral.

In the Generic Access Profiles (GAP), three modes are defined:

- Mode 1: none. No security function is activated, all Bluetooth devices may be freely connected.

- Mode 2: application-level security. This mode guarantees a security after connection establishment. This mode supports various controls according to applications and associated functions act at application level.

- Mode 3: link-level security. It is the most secure mode: it inherits from mode 2 functionalities by adding a preliminary control (authentication, encryption) at the time of connection attempt.

Security level 2 applies identification procedure when accessing services. Level 3 is more strict, it runs authentication procedure during the connection establishment. Following a successful authentication, keys are generated before establishing a protected encrypted communication.

In addition to these three security modes, Bluetooth offers two security levels for physical devices (reliable or not reliable). Trust levels are defined. Peripherals can have two trust values: trusted, untrusted. A trusted peripheral can access all services without restriction. The peripheral must be authenticated. The untrusted peripheral has only limited services. Every new peripheral is considered as unknown and is untrusted.

Link Layer Security. In each Bluetooth peripheral, four elements are used to maintain link-level security:

- The permanent Bluetooth peripheral address BD_ADDR of 48 bits, unique for each peripheral, it is assigned during its construction. These addresses are controlled by the IEEE and can be regarded as Ethernet MAC addresses.

- A four-digit personal identification number (PIN) code, it is a personal code assigned to the user. It can be stored on 1 to 16 bytes.

- A private authentication key *"link key"* of 128 bits.

- A private encryption key with size varying from 8 to 128 bits.

- A random number generator RAND of 128 bits.

BD_ADDR addresses of other peripherals are stored in peripheral database for later use. All peripherals in a network are synchronized. Communication between various peripherals uses a fast FHSS technique which, in addition to its robustness to interferences, makes difficult to capture transmissions.

Security of Services. Needs for authorization, authentication and encryption change. At the moment of connection establishment, the user can choose among several security levels. The security level is defined by three attributes:

- Requested authorization: access is automatically granted to trusted peripherals or to untrusted peripherals after an authorization procedure.

- Requested authentication: the peripheral must be authenticated before connection to the application.

- Requested ciphering: the link must be in an encrypted mode before accessing services.

At lower levels, services can be fixed as being accessible to all peripherals. Usually, the restriction urges the user to be authenticated. When highest security levels are necessary, both authorization and authentication may be needed.

A reliable peripheral has access to services whereas an unreliable peripheral needs an authorization.

Bluetooth offers three security levels:

- Authorized and authenticated

- Authenticated

- Free access

Key Management. Several types of keys exist. The main key is the link key used between two peripherals for authentication purpose. From this key, the encryption key is derived to secure data. It is regenerated at each new transmission.

Authentication Key *link key*. According to applications, there are four types of keys of 128 bits length, temporary (used only during the current session) or semi-permanent (used in several sessions):

- Initialization key K_{init}

- Unit key K_A

- Combination key K_{AB}

- Master temporary key K_{master} (master key)

Each one of these keys is generated with an algorithm of E family. Except for E0, all algorithms are based on Secure And Fast Encryption Routine (SAFER+) [MUL 01]. It is a simple and robust algorithm to linear or differential cryptanalysis.

Link key is used during all security exchanges. As soon as a link key is generated, its authenticity is checked.

A short description of each one of these keys and procedures of their generation are given in the following sections.

Initialization key. The initialization key K_{init} is used during initialization phase. This key protects initialization parameters. It is used for only one session and is created at each device initialization. It is used when there is no yet unit key or combination key.

Unit key. The unit key K_A is derived by the equipment during the installation phase. K_A storage requires a limited memory capacity. It is semi-permanent and is stored in the volatile memory.

Combination key. The combination key K_{AB} is derived by two peripherals A and B to start a communication. This key is generated for each pair of peripherals and is used when security must be optimized. More memory is required since the peripheral needs to store a combination key for each connection. This key often replaces the unit key for a certain period.

Master key. The Master key K_{master}, is temporary and is used when the main peripheral wishes to communicate with several slaves peripherals (point-to-multipoint communication). It replaces the current link key.

Encryption Key. The encryption key is derived from the current authentication key. It can be shortened because of reglementory constraints.

Although it does not constitute a key, a PIN code is used to identify a peripheral.

PIN Code. It represents a number which can be fixed or chosen by the user. Its size is usually of 4 bytes but it can be on 1–16 bytes.

The user can change it when it wishes it in order to increase security. This PIN code can be introduced into only one peripheral (fixed PIN) but it is more secure to introduce it into several peripherals.

All in all, the main security parameters are shown in Figure 6.21.

E_{22}: generation algorithm of link keys
E_3: generation algorithm of encryption keys

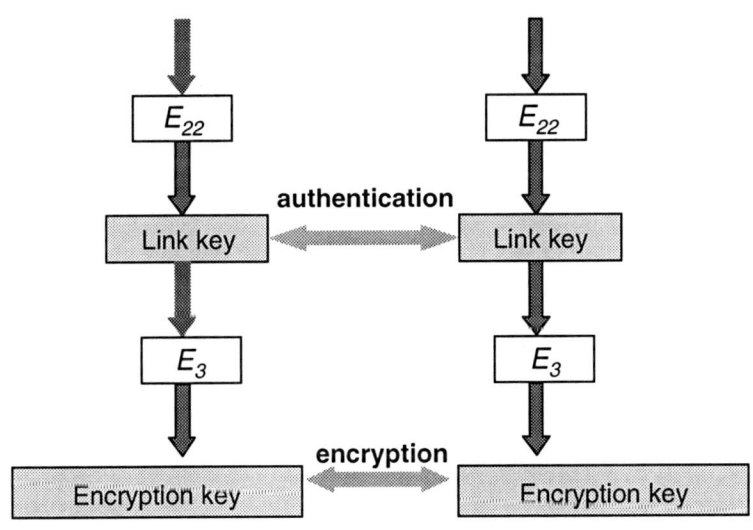

Fig. 6.21. Security elements

2.2 Main Procedures

The initialization phase represents an important phase. It includes four tasks which are:

- Generation of the initialization key
- Authentication and generation of the authentication key
- Exchange of link key
- Generation of encryption key

Obviously, execution of these various tasks depends on the required service. Security manager fixes the security level. If necessary, it can launch authentication operations and/or encryption and can send requests to user interface to control authorization and PIN code.

Communications between various Bluetooth entities use a link key. This key of 128 bits is generated in a different way according to the communication type:

- At initialization step (unit key)
- During the starting of a communication between two peripherals (combination key)
- During a communication initiated by only one device (initialization key),
- During a communication between a master and several slaves (master key)

A total diagram describing the various phases is given by Figure 6.22.

We assumed that only two peripherals A and B wish to communicate. The following steps occur:

- A and B switch on (*First Startup*)
- A and B communicate in order to establish a shared secret key for next communications (*First Handshake*);
- A and B use the non-volatile stored key to exchange data (*Following Handshakes*)

The operation *First Hansdshake* is launched when two units wish to communicate for the first time. One of the two units (requestor unit B) tries to communicate with the other unit (verficator unit A). B must show that it is authorized to communicate. By definition, each requestor unit is authorized with the condition it shares the same code PIN with the verificator.

Security in WLAN, WPAN, WSN and WMAN

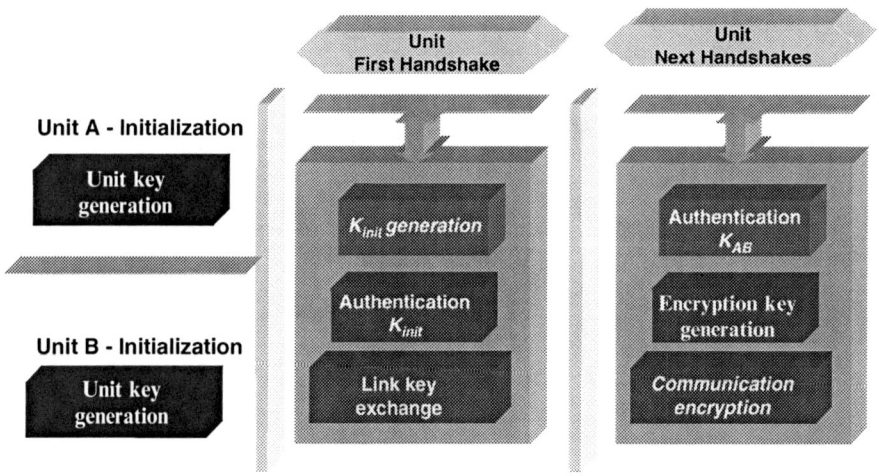

Fig. 6.22. Global security scheme

Generation of Keys. Bluetooth handles a whole of keys. The generation of these keys is a delicate process. In this part, we give a description of the procedure of generation of each of the four keys listed previously.

Initialization Key Generation. This key is necessary when two units launch for the first time a communication. The PIN code is introduced into the two peripherals. The algorithm used for the generation of initialization key is called E_{22}, it is held according to following steps:

- Unit A sends a random number IN_RAND to unit B.

- The two units generate the initialization key $K_{\text{init}} = E_{22}$ (PIN, PIN_LENGTH, IN_RAND), it is used like a temporary link key (Figure 6.23).

The success of this operation depends on the agreement of the two codes PIN and PIN'.
The same algorithm is used for the generation of the master key.

Unit Key Generation. This operation is carried out during the first powering of the peripheral. This last generates itself its unit key (K_A) based on its address BD_ADDR$_A$ and a random number RAND$_A$. This key is stored in the non-volatile memory. The algorithm used for the generation of this key is called E_{21} (Figure 6.24).

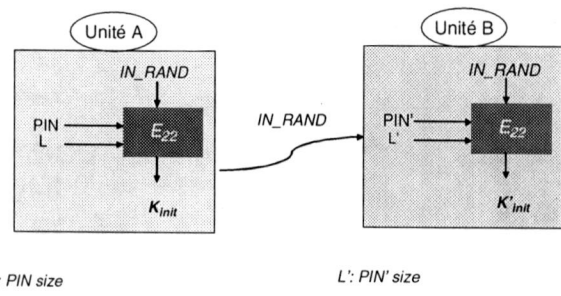

Fig. 6.23. Initialization key generation

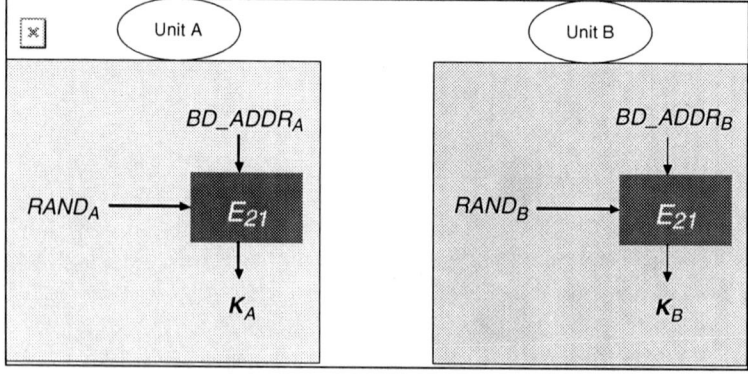

Fig. 6.24. Unit key generation

This key may be rejected or regenerated, consequently all the links which this unit become unavailable and the *First Handshake* phase is reiterated with each equipment with which it communicates.

***Master* Key Generation.** This key is generated by the master by using the E_{22} algorithm with two 128-bit random numbers; the output is a 128-bit key. A third random number is then sent to the slave station. With the algorithm and the current link key, an overlay (OVL) is calculated by master and slave units. The new link key which is the key master is thus sent to the unit XORed with OVL. The slave unit can thus determine the key master. This procedure is carried out with each slave.

Combination Key Generation (see section 2.2.2).

Encryption Key Generation. The encryption key is generated using E_3 algorithm. The procedure is described by Figure 6.25.

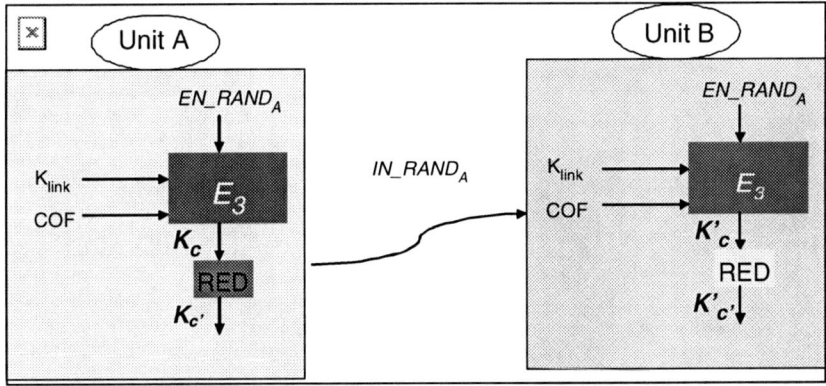

Fig. 6.25. Encryption key generation

The encryption key noted K_c is generated from the link key, a random number generated by A and a parameter 96-bit ciphering offset (COF). COF depends on the type of the communication in progress. It can be:

- Generated during the phase of authentication challenge/response, it is called ACO (authenticated COF)
- The address of the transmitter

If necessary, the encryption key can be shortened by using RED function. The key is changed automatically at the beginning of each encryption operation. It is generated at each time the link manager (LM) activates ciphering.

Each Bluetooth peripheral has a generator of random numbers used in authentication/encryption operations.

Authentication. Bluetooth uses a challenge/response mechanism with shared symmetric keys. The authentication begins by emitting a request to another peripheral by exchanging BD_ADDR and link key. Upon authentication, encryption is used to communicate. Without knowledge of PIN code, a peripheral cannot be recorded if the phase of authentication is activated. In order to facilitate the procedure, the PIN code can be stored inside the unit (e.g. in a memory or a hard disk).

Bluetooth uses a challenge/response exchange for authentication in which the knowledge of the requestor of the secret link key K_{AB} (which is K_{init} at first authentication operation) is controlled through a protocol applying symmetric keys. A temporary link key is used at the time of the first contact, thereafter the semi-permanent key shared by the two units A and B is used. The iterative per-block symmetric encryption algorithm E_1 is used.

This protocol is represented in Figure 6.26.

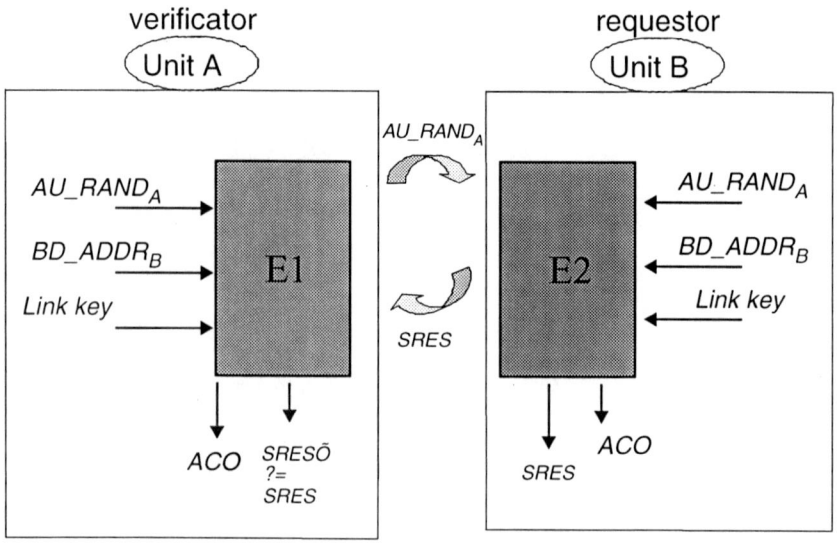

Fig. 6.26. Authentication procedure

Unit A sends a random number, noted AU_RAND_A, with a code of authentication noted E_1 for the unit B. The unit B calculates SRES and returns the result to unit A as shown in Figure 6.26. Unit A calculates SRES′ and will authenticate the unit B if SRES = SRES′. SRES is calculated by using algorithm SAFER + based on the E_1 function which takes as input parameters (AU_RAND_A, address BD_ADDR of the peripheral, current link key). At each authentication a new random number AU_RAND_A is generated.

The link key can be the temporary initialization key K_{init} or the effective link key between A and B K_{AB}. The success of this procedure depends on the equality of the two link keys of the two peripherals.

In addition to the result being used to authenticate B, the E_1 function produces a number ACO that will be used in cipherin/deciphering computations.

The application indicates the unit which must be authenticated and by which element of the network. Some applications require that an authentication takes place in only one direction. However, in some point-to-point communications, a mutual authentication may be preferred. The link manager coordinates the authentication preferences required by the application to determine in which directions the authentication must take place. In case of failure, authentication operation is reiterated during a certain time.

Exchange of Keys. The exchange of keys can take place during initialization phase carried out separately by each of the two peripherals which wish to set up an authentication/encryption phase.

The link key K_{AB} must be exchanged between the two units A and B. The choice of this key depends on the required security level and resource constraints (memory capacity) of the two communicating units. When a protected authentication is necessary, the link key is built based on the combination key. We distinguish two cases.

First Case. When the link key is the key of the unit, it is transmitted from one unit to the other by carrying out simply an XOR operation with the initialization key (Figure 6.27).

Second Case. Let us consider the link between combination and link keys (see Figure 6.28).

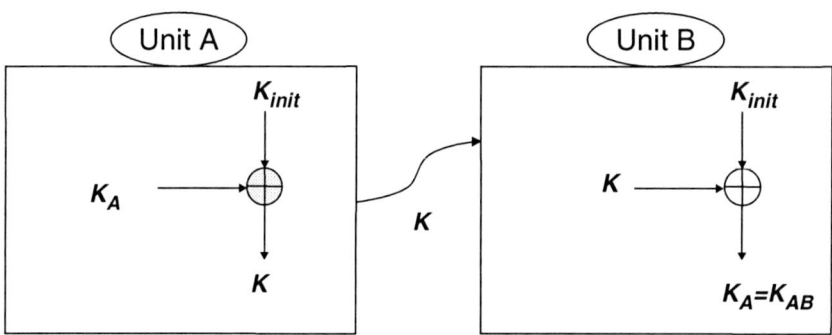

Fig. 6.27. Unit key generation

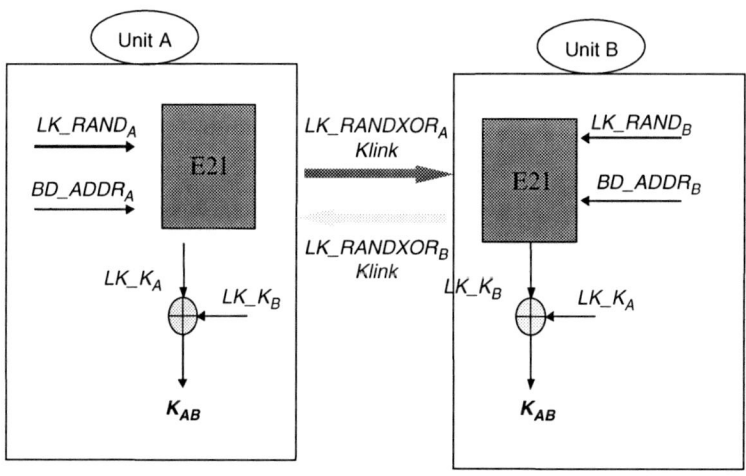

Fig. 6.28. Generation of combination key

When the link key is the combination key, the following procedure is carried out:

- Unit A generates a random number LK_RAND_A and calculates (by using the same algorithm to generate the new key of the unit) $LK\_K_A = E_{21}(LK\_RAND_A, BD\_ADDR_A)$.

- The unit B generates a random number LK_RAND_B and calculates $LK\_K_B = E_{21}(LK\_RAND_B, BD\_ADDR_B)$.

- LK_RAND_A is sent to B and vice versa.

- Each unit calculates the key and the current combination key K_{AB} is a simple XOR between the two keys.

Once the link key is exchanged, the initialization key is not valid any more and a new procedure of authentication by using the new semi-permanent key is necessary. The encryption key is derived from this new link key.

Encryption. Useful information part (payload), in a Bluetooth packet, can be encrypted whereas access code and header cannot. Specification 1.0 describes an encryption algorithm linear feedback shift registers (LFSR). The size of a register is of 128 bits and the effective size of the key can be selected between 8 and 128 bits. This variation of size allows Bluetooth to be used in countries having regulation limitations. If a combination key or a unit key is used, broadcast traffic is not encrypted. Point-to-point traffic can be encrypted or not. There are several encryption modes related to the type of the key (semi-permanent link key or a master key). When the master key is used, three modes are possible:

- Encryption mode 1: none.

- Encryption mode 2: the point-to-multipoint traffic is not encrypted but the point-to-point traffic is encrypted with the master.

- Encryption mode 3: all the traffic is encrypted with the master key.

Key size must be negotiated between master and slave units before starting communications. In each peripheral, there is a parameter which contains the authorized maximum key size. In the same way, in each application, there is a parameter of the acceptable minimal key. The application can thus fail the negotiation and encryption can not be carried out (Figure 6.29).

Encryption/decryption mechanism is rather close to that of authentication. The difference concerns parameters and used algorithm. In Figure 6.27, it is seen that the algorithm generates a key noted K_{cipher}. Parameters are: a random number EN_RAND_A, BD_ADDR_A, clock of A ($clock_A$) as well as a key K_c.

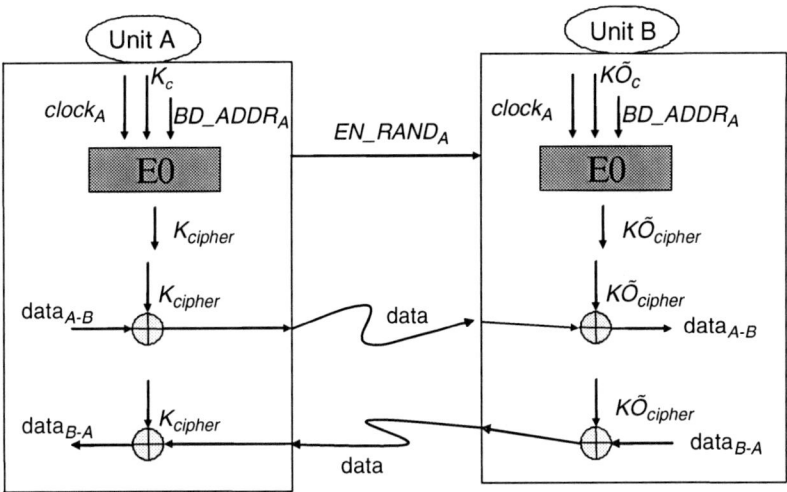

Fig. 6.29. Encryption scheme

The K_c key is created by both A and B by using a random number, ACO previously calculated at the time of authentication and the link key.

Security in Bluetooth is adapted to relatively small networks. If the requirements in security are stronger, one will need to use more robust algorithms.

2.3 Security Flaws

We examine in this section a list of vulnerabilities of Bluetooth security systems [HAG 03].

Denial of Service. One of the faults of Bluetooth protocol and ad hoc networks in general is a fault specific to wireless networks: DoS per battery. As elements of a network are mobile, they are dependent on a limited energy source. Energy-saving mechanisms are implemented. One of the principal DoS attack is to overload a device targeted so as to exhaust its batteries as soon as possible. If an attacker can damage or destabilize energy-saving mechanisms the whole network can be seriously affected because several functionalities hardware and software can be degraded such as frequency hopping. Within Bluetooth SIG, work is undertaken to alleviate this drawback.

Confidentiality. Bluetooth being a wireless network, every one can listen data frames emitted on wireless links. It is thus rather easy to identify the protagonists using PIN code or Bluetooth address of the physical devices for established connections. Confidentiality issues are then. The encryption key element is the link key (random numbers are transmitted in clear); knowing this

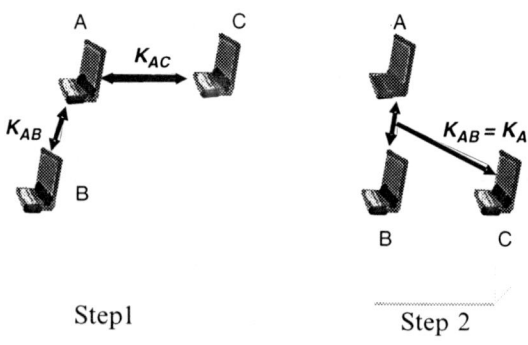

Fig. 6.30. Stealing of unit key

key and listening during the authentication phase are sufficient to decipher the communication.

Identity Stealing. Figure 6.30 presents the steps of an identity stealing.

During the use of a link key based on unit key, it is rather easy to steal the identity of a correspondent. By assuming that A and B communicate based on unit key of A, a third speaker C can come to communicate with A and to obtain this key. C can thus use the Bluetooth address of B to beconsiderd as him. If A and B are communicating, C will need the encryption key. To generate it, it needs the unit key of A, ACO, generated at the time of the authentication phase between A and B and the exchanged random number. If C has the unit key, it then just listens to the initial communication between A and B because ACO and the random number is exchanged in clear meanwhile this phase.

Constraining Security. During the establishment of a connection between two parts, the PIN code of the user must be stored on both devices. This action is very constraining for the user who will tend to withdraw this functionality.

Algorithms Security Gaps. The principal weakness of Bluetooth security is the insufficiency of the personal code. The PIN code has a considerable importance since it is used in combination with other parameters to generate the link and encryption keys. By using a too short code, proprietary algorithms are relatively ineffective. It is rather easy to guess this code if this one does not exceed four digits (standard length in the applications of Bluetooth). Moreover, the same PIN code must be introduced into the units to initialize a protected communication. This introduction poses a critical security problem.

Another problem is related to the unit key. Authentication and encryption are based on a secret shared link key but need other information which are transmitted in clear.

Frequency Hopping Sequence Attack. In a large majority of wireless systems, omnidirectional propagation is used. Bluetooth peripherals are considered as relays points to extend the communication zone but this involves attacks. An attacker can easily intercept the traffic which crosses (the windows, walls). From there, it will be able after a certain period, to collect relevant information on the operation of the network allowing him to discover the frequency hopping algorithm as well as its parameters.

BD_ADDR Spoofing. The address of a peripheral is single. Nevertheless, this unicity involves another problem. Once, this identification is associated to user, peripherals can thus be tracked and spied.

2.4 Synthesis

Bluetooth security system is flexible thanks to its several security modes and a series of functionalities which are:

- The exchange and the regeneration of keys

- Authentication driven by the application (authentication in only one direction or mutual authentication):

- Encryption:

    - With keys of 8 bits to 128 bits
    - With modified algorithm SAFER+
    - From station to station

- Two main key types:

- Authentication key (link key) which can be:

    - An initialization key
    - A temporary key (the master key)
    - Unit key
    - A combination key

- An encryption key.

Bluetooth devices use a combination of PIN code and a physical address. In order to know the identity of the user, peripherals and their rights, concepts of authorization and authentication are defined. From the authentication point of view, the peripheral is only authenticated and not the user. This point is critical for the applications requiring a strong security. Bluetooth technology uses a link layer security where a secret key is allocated to each connection and an encryption key is derived from authentication key. The access is allowed only after connection establishment. Access control can be asymmetric but once a connection is established the data flow is typically bidirectional.

3. ZigBee Security

ZigBee is designed as a global hardware and software standard for wireless networking devices. Its main features are: highly reliable, low cost, low power, low data rates and highly secure.

Three security levels are specified: none, ACLs and symmetric key employing AES 128-bit encryption. The concept of "trust centre" is used. We can use link and network keys. Two operations are supported: authentication and encryption. Security can be customized for applications and keys can be hardwired into applications.

The main elements of security mechanisms provided in ZigBee are summarized in Figure 6.31.

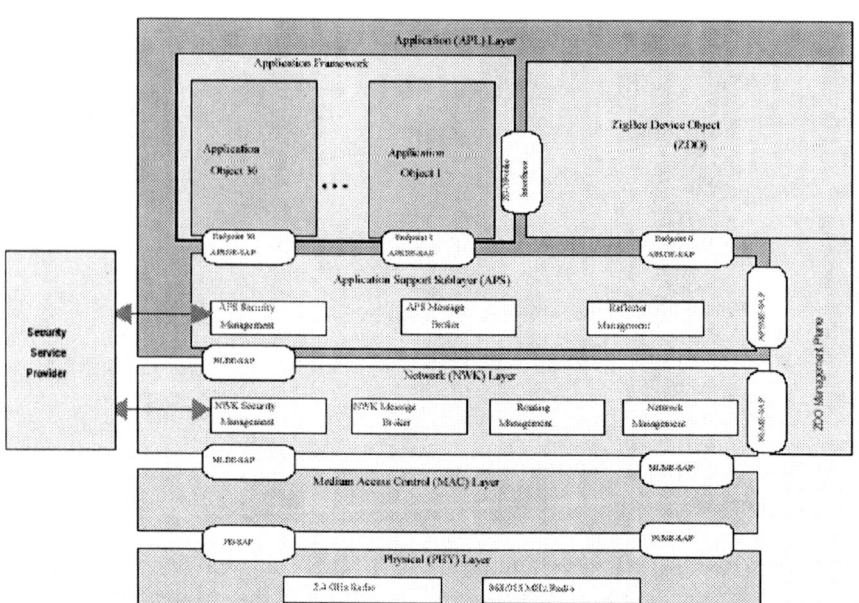

Fig. 6.31. ZigBee security architecture

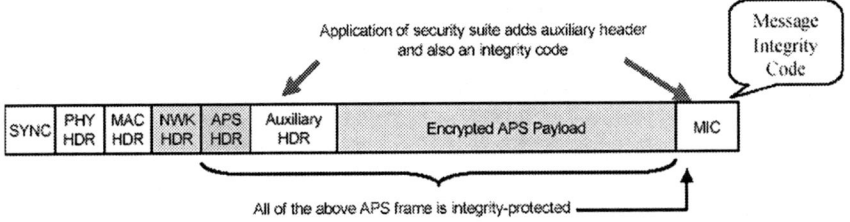

Fig. 6.32. Frame structure

3.1 ZigBee Security Architecture

Different types of keys are defined: master key, link key and network key.

The master key, designed for long-term security between two devices, can be set up over-the-air or by using out-of-band mechanisms (eavesdropping should be prevented during this setup phase). It is sent by the trust centre. It can also be a factory-installed option.

The link key is provided for security between two devices. It is derived from the master key. It can also be factory-installed option.

The network key serves to provide security across the network and protects against outsiders attacks. It can also be factory-installed option.

Link and network keys can be updated periodically. These keys need to be set up with and between new devices that join the network. If keys are set up over-the-air only, the last link is vulnerable to a one time eavesdrop attack. After a device joins it needs to store multiple keys. Policy decisions are not defined in the current standard, it may include, for example, out-of-band methods for key set-up, cost/security tradeoff for number of link keys needed, selection of commercial/residential modes, security error conditions handling, loss of counter synchronization support, loss of key synchronization, policy for expiration and update of keys, policy for accepting new devices, etc.

The structure of the frame is given in Figure 6.32.

Several headers can be added to data frames at the MAC, network and application layers.

3.2 Mechanisms

Freshness. Freshness check prevents replay attacks (an attacker from replaying messages). ZigBee devices maintain incoming and outgoing freshness counters. Counter is reset when a new key is created. Devices that communicate once per second will not overflow their freshness counters for 136 years.

Message Integrity. Message integrity prevents an attacker from modifying the message in transit. We can use 0, 32, 64 or 128 bits for integrity check. The default value is 64. Integrity options allow tradeoff between message protection and message overhead.

Authentication. Authentication provides assurance about the originator of the message. It prevents an attacker from modifying a hacked device to impersonate another device. Authentication is possible at network level or device level. Network-level authentication is achieved by using a common network key. This prevents outsider attacks while adding very little in memory cost. Device-level authentication is achieved by using unique link keys between pairs of devices: This prevents insider and outsider attacks but has higher memory cost.

Encryption. Prevents an eavesdropper from listening to messages. ZigBee uses 128-bit AES encryption. Encryption protection is possible at network level or device level. Network-level encryption is achieved by using a common network key. This prevents outsider attacks while adding very little in memory cost. Device-level encryption is achieved by using unique link keys between pairs of devices. This prevents insider and outsider attacks but has higher memory cost. Encryption can be turned off without impacting freshness, integrity or authentication. Some applications may not need encryption protection.

3.3   Trust Centre Concept

The trust centre allows devices to join the network and distributes keys. The ZigBee coordinator is assumed to be the trust centre but it is possible for the trust centre to be a dedicated device, e.g. a portable device.

As shown in Figure 6.33, a typical network is composed of a coordinator, routers and end devices.

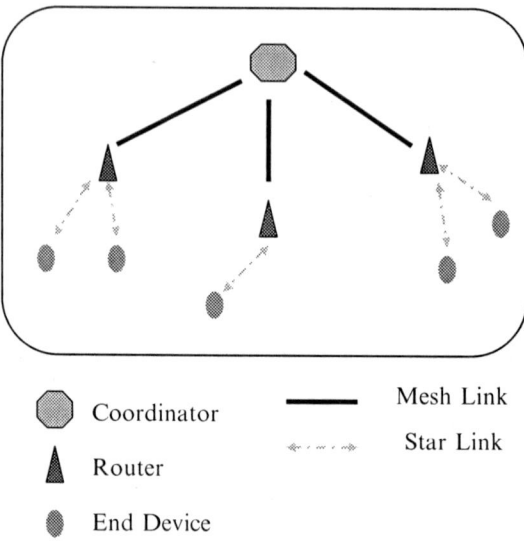

Fig. 6.33. Typical network

The coordinator (trust centre) may have several roles and modes.
Trust centre roles
We can find three roles:

- Trust manager: it authenticates device that request to join network

- Network manager: it maintains and distributes network keys

- Configuration manager: it enables end-to-end security between devices

Trust centre modes
Two modes are defined.

- Residential mode: the trust centre allows devices to join the network, but does not establish keys with network devices. The trust centre cannot update keys periodically because it does not maintain keys with network devices. The memory cost in the trust centre is minimal and does not scale with the size of the network.

- Commercial mode: the trust centre establishes and maintains keys and freshness counters with every device in the network. This allows centralized control and update of keys. Cost memory in the trust centre could scale with the size of the network.

4. WiMAX and IEEE 802.16 Security

4.1 IEEE 802.16 MAC Security Sublayer

The security is considered as a sublayer and implements privacy, authentication and confidentiality for the wireless network. Like IEEE 802.11 it is restricted to the link so it concerns connections between the base station and stations.

Since IEEE 802.16 is probably not used in a private context but for an operator willing to earn benefits from this service, the security system is normally robust against theft of service such as bandwidth, connections and so on. The base station is also protected against management attacks. Since the system is strongly centralized a certificate-based security is very suitable and easy to deploy.

The security service is hence divided into two parts: the authentication/key derivation and the encryption (Figure 6.34).

Like IPsec, IEEE 802.16 defines the security association (SA) that the base station and stations hold to support the application of security. They define *Primary*, *Static* and *Dynamic* SAs. Stations must have a primary SA at initialization. The static SA is only concerning base stations. Dynamic SAs are created and removed at the speed connections are created and removed. Static and dynamic SAs can be shared by several stations if the connection is a group communication.

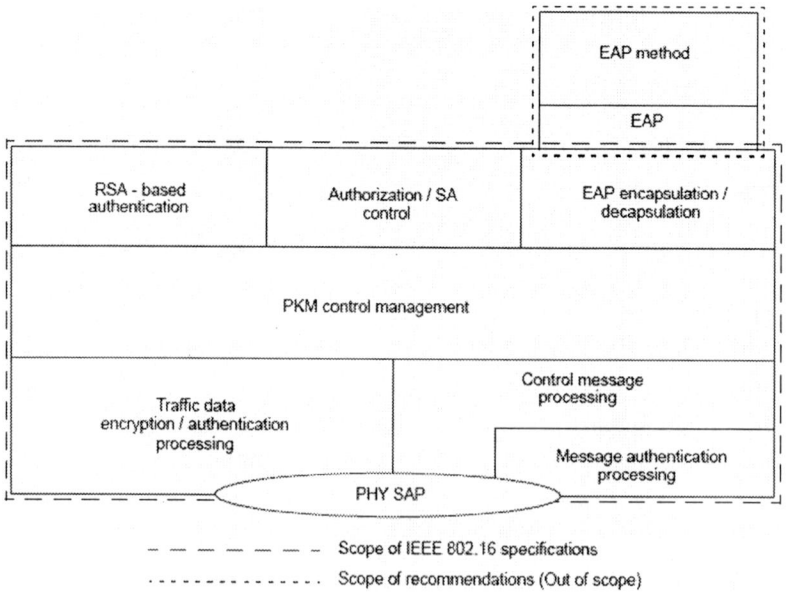

Fig. 6.34. Security services

Of course, in the SA we will find the negotiated cryptographic suite, its identifier called an SAID. The primary SAID is to manage the main connections with the base station and its SAID is equal to the primary CID. This SA is also responsible for storing keying material with its lifetime so that whenever an expiry happens a rekeying happens between the station and its base. The defined suites in the IEEE 802.16 2004 define encryption, authentication and method for encryption, they were as follow:

- No encryption and no authentication, 3DES 128
- CBC 56-bit DES, no authentication and 3DES
- No encryption, no authentication and RSA 1024
- CBC 56-bit DES, no authentication and RSA 1024
- CCM AES, no authentication and AES 128

4.2 PKM Protocol

Privacy key management protocol is responsible of distributing keys between BS and the station. It is also used as an access control mechanism. The protocol has two versions. The first is defined in the fixed network and the second version is added into the mobile "E" addendum.

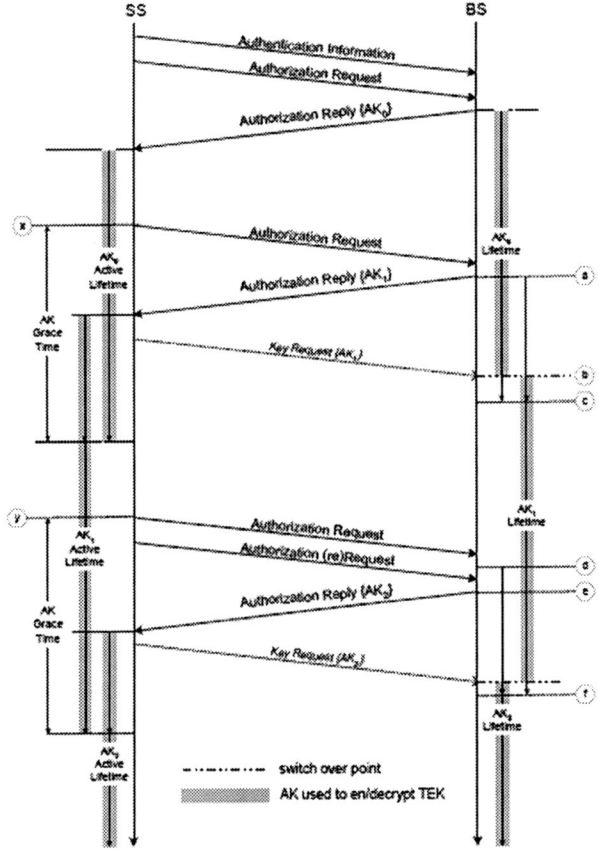

Fig. 6.35. PKM initialization procedures

PKMv1. Messages sent for authorization and data encryption key derivation are under the general name of PKM messages.

The station sends an *Authentication Information* message to the base station. The message contains the station certificate. The station sends then *Authorization Request* message that asks for an authorization key. In this message there is also:

- The crypto algorithms supported: authentication and encryption

- CID, the primary one

The authorization phase in the standard is simply an authentication based on certificates. The initiation happens from the station side. A set of messages are defined for that purpose such as:

- Authentication request

- Authentication reply
- Authentication reject
- Authentication invalid
- Authentication information

We will not go through the standard defined state machine since things are straight forward and follow a classical authentication. We summarize only the main idea in the authorization phase hereafter. This is summarized in Figure 6.36.

At the beginning the station sends the information request and the registration request. The registration request message contains an X.509 certificate with the station public key digitally signed by another certification authority (CA). The request contains also the available crypto parameters in the station. All these messages are HMAC signed with the SHA-1 algorithm and RSA crypted.

The reply is signed and crypted by the public key of the station, it contains:

- An AUTH-Key (AK) encrypted with the station public key
- The AK active lifetime
- Key-Sequence-Number AK sequence number

Normally it is known that a sequence number would prevent from replay attacks. Further on any security-related message should be signed by the AK

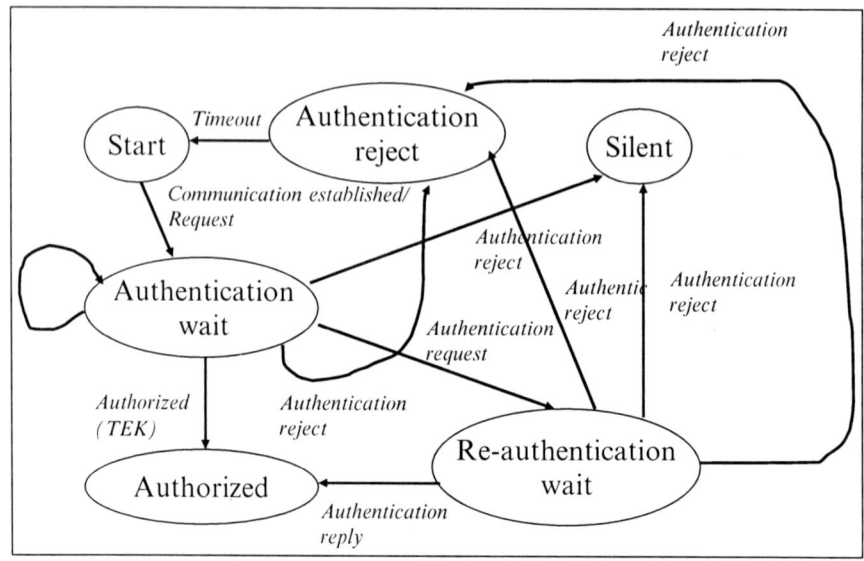

Fig. 6.36. Authorization state machine

and we note that there are two AKs for each direction of data flow. So in short the AK serves as follows:

- To check the HMAC in key request going from the station to the base station
- To calculate the HMAC in the key reply
- To encrypt the TEK in the key reply from base station to the station.

4.3   Traffic Encryption Key

After authorization a state machine for each SAID is started, mainly to derive the data encryption key. The state machines have to periodically refresh the encryption key so they use a special message: *Key Request* to the base station, with a reference to the actual SAID.

The base station responds with a *Key Reply* message, containing the new key for that particular SAID. TEK messages are encrypted using a derivation of the AK called *key encryption key* (KEK). In the TEK reply the base station sends the necessary initialization parameters relative to the encryption (e.g. CBC IVs).

4.4   Security Enhancement for Mobile Communications

The new features are mainly related to three things:

- EAP integration in the authentication phase
- A version two PKM algorithm
- Group key derivation
- Another integrity check method called Cypher-based MAC
- Mention of pre-authentication but no proposals yet

We know that EAP opens the door to an infinite number of authentication methods including 3G like methods. So it is not anymore restricted to PKI systems.

As for the pre-authentication, it simply means that a node should derive the AK before it moves to another base station. How this is done and when is not really explained and is left for any eventual next version of the standard.

We explain hereafter the PKMv2. It uses a slightly different key hierarchy that defines what keys are necessary and how they are generated. It is different due to the presence of two authentication systems, one based on RSA and the other on EAP.

The first keys are used to protect management message integrity and transport the traffic encryption keys (TEK). The TEK are derived through the first keys

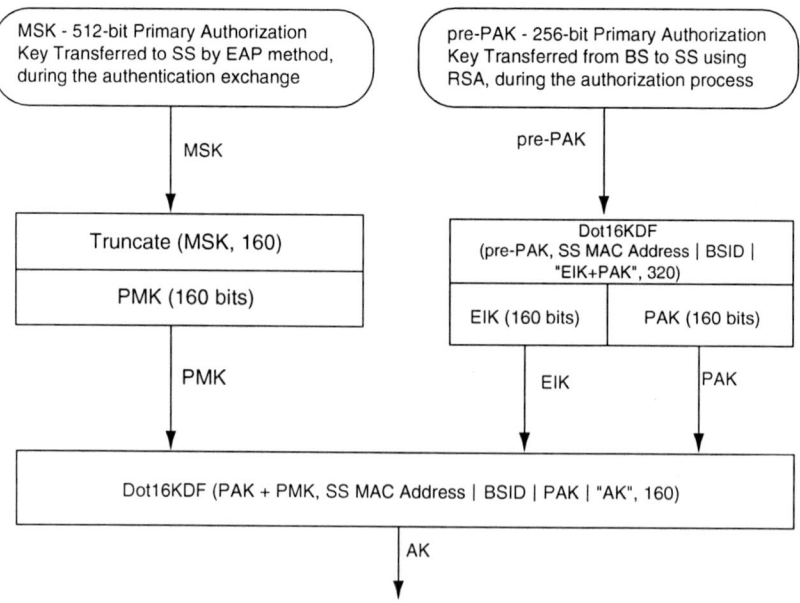

Fig. 6.37. PKMv2 used in mobile 802.16

plus as usual some exchanged parameters in the initialization. They rename the RSA-generated keys as the pre-primary AK (pre-PAK), and the ones generated through EAP-based authentication as the

The mobile station keys MSK where MSK is a shared "master key" derived by two peers.

The general procedure to extract AK from those is as follows:

If (PAK and PMK)
AK ⇐ Dot16KDF (PAK PMK, SS MAC Address | BSID | PAK | "AK", 160)
Else If (PMK and PMK2)
AK ⇐ Dot16KDF (PMK PMK2, SS MAC Address | BSID | "AK", 160)
Else
If (PAK)
AK ⇐ Dot16KDF (PAK, SS MAC Address | BSID | PAK | "AK", 160)
Else // PMK only
AK ⇐ Dot16KDF(PMK, SS MAC Address | BSID | "AK", 160);
Endif
Endif

So we can see that it is very close to the IEEE 802.11 four-way handshake except that it is called here a three-way handshake where one handshake is removed because everything is centralized to the base station and there is no need to ask for nonces coming from the station.

5. Conclusion

Security in wireless networks poses new problems not encountered in fixed networks. Indeed, flexibility in displacements of users makes security control very complex. In addition, because of wireless transmission characteristics, a huge number of vulnerabilities appear. Up to date, no communication channel is as variable or unverifiable as a wireless link. In order to reach a satisfactory security on such networks, it is necessary to know their intrinsic faults:

- Broadcast of information make easy to execute remotely passive attacks
- Sensitivity to jamming involving a DoS and decreasing the availability of the network
- Not secured configurations by default of new devices
- Lack key distribution mechanisms at link level, etc.

The strong polemic around IEEE 802.11 security issues, gave a great dash to research on wireless security topic. Indeed, an intense activity was undertaken as well in the field of cryptography as in the field of architectures of authentication. Incidences of technological improvements through proposed solutions have and will have a great impact on both wireless and wired networks.

For Wi-Fi technology, an important list of security faults was detected because mainly of the weaknesses of protocol WEP and the absence of mechanism of key distribution. Hopefully, the standard IEEE 802.11i specified powerful mechanisms to enhance security. However, some attacks are still possible like dictionary and DoS attacks.

Security procedures in Bluetooth are more elaborate than in Wi-Fi, but there are always weaknesses:

- A simple authentication where only peripherals are authenticated, no authentication of users nor data
- Significant information exchanged in clear during security operations
- Security specifications which consider only simple problems
- A lack of dynamic keys

Authorization system must be reinforced. This is possible without changing the protocol stack but by modifying the security manager and the initialization procedure. Other improvements are possible:

- Exchange keys between peripherals by using key agreement-type procedures

- Frequent change of the pseudonym address which masks the peripheral physical address

- More robust methods of security must be implemented in high layers

- Reinforcement of the system of authorization with architectures of distribution and management of keys

- Secure routing for broader networks must be investigated separately

ZigBee provides pretty good security mechanisms for low rate and short range wireless systems. These mechanisms are settled in combination between layer 2 and high layers. Revocation is not supported but next versions of the standard will probably include it. WiMAX security applies a classical approach based on using certificates. Also, the use of EAP makes WiMAX security similar to the one adopted in Wi-Fi networks. Beyond the formation and sensitizing of users, it is essential to configure its network in a protected way. This step includes configuration of link and transport layers but also periodic audit and permanent monitoring of its network.

Chapter 7

Practising

In this chapter we are going to get into some practical details of technologies we have explored and compared along the book, with a special focus on wireless sensor networks (WSN) experiments and WiMAX deployment.

Thus, our aim is to show by examples how to start and prepare try-outs for WSNs and WMANs.

For our WSNs test cases, the material is organized giving a brief overview of TinyOS, an open source project devoted to network embedded technology and devices. Then some experiments and results are presented in a step-by-step fashion, to provide the interested reader suggestions and feeling with WSNs.

As for WMANs, the main focus is on configuration and testing of a complete wireless system composed of a base station and a couple of subscriber units using fresh certified WiMAX equipment. A lab set-up for experimenting basic network performance is showed, together with results attained, by using popular tools.

1. Mastering the TinyOS Platform

The main objective of this section is to introduce the WSN platform used for our experiments. Then we will move on to practical aspects and details related to Mica-Z devices (or motes) and test cases in three typical indoor deployment scenarios. For each of these we describe context, methodology, tools and final results. We are interested in sample topologies to test basic WSN operation and functions. References to inspiring work and available papers on the Internet are given along for further readings and information.

1.1   Brief Historical Notes and State-of-the-art in TinyOS

TinyOS is an open source operating system developed by the Wireless Embedded Systems (WEBS) Research Centre at Berkeley University. Following links from its main site, that is http://webs.cs.berkeley.edu/, we end up to a group of projects and sub-projects devoted to this field, the most articulated is NEST, an acronym for Network Embedded Systems Technology.

The objective of the NEST project is to make available an open experimental software/hardware platform for NEST research, in order to accelerate the development of algorithms, services, and their composition into modular complex applications.

To better understand TinyOS origins, below are reported with comments key elements of the composed platform pursued by NEST, which consist of:

- The hardware required for low-cost, large-scale experimentation of Network Embedded Systems, to be based on small, resource-limited devices (sensor nodes)

- The specialized operating system that supports applications, as well as a set of features tailored to the context, such as communication, low-power consumption for maximum battery lifetime, remote monitoring and control

- A debugging and visualization environment, suited to operate with numerous interacting nodes and to support a development model based on events

- The infrastructure services for time synchronization, storage, computing and simulations, the last being powerful enough to explore large networks and worst-case situations like those encountered in wireless communication

- Mechanisms for composition of finite-state machines that enable modular design and help in developing stable applications

- A macrocomputing language that simplifies programming groups or collection of nodes, that is usually the case working in a typical distributed environment

The focal concept around which the NEST platform is developed seems to be algorithms and algorithmic work, allowing researchers to move from theory to practice, without each group developing extensive infrastructure for testbeds and simulations.

A list of the main projects under the NEST umbrella started after its inception, including the fundamental TinyOS, is reported for further reference and to show different lines of development:

- TinyOS: Operating system support for tiny networked sensors
- NesC: custom compiler support for TinyOS
- TOSSIM: TinyOS mote simulator
- TinyDB: query processing system of TinyOS sensor networks
- TinySEC: link-layer encryption for tiny devices
- Maté: virtual machine for TinyOS motes
- FPS: network protocol for radio power scheduling in WSNs

It is easy to match this list with key elements just seen for the NEST platform. Other special sub-projects are aimed to specific purposes: *Ivy*, a sensor-network infrastructure for Berkeley's College of Engineering; *Calamari*, localization solutions for sensor networks; *Great Duck Island*, employing motes for habitat monitoring in sensitive wildlife habitats; *CotsBots*, designing inexpensive and modular robots from off-the-shelf components.

Looking at popular subjects discussed on TinyOS mailing lists, one of the main sources of information for the community of users and developers, software issues regarding TinyOS itself, nesC programming and TOSSIM are prominent, together with specific questions on two widely used hardware platforms, manufactured by Crossbow and Moteiv.

There is a great deal of interest around this open source project, with training sessions and presentations coming up on a regular basis, and international events for researchers, developers, sensor node and chip manufacturers, such as TeleTeXt (TTX), TinyOS technology exchange, at its third edition in 2006.

That has led to a huge effort to make it available to several platforms[1] and numerous sensor boards. From a numerical point of view, the success of the project is in the wide community using it to develop and test various algorithms and protocols.[2] For those interested in more statistics, the project page at http://sourceforge.net/projects/tinyos/, registered in June 2001, has a complete overview of TinyOS activity and trends.

Currently, the latest version of TinyOS is 1.1.15, released on December 2005. There is an internal working group[3], which developed a completely new version 2.0 of TinyOS, at its beta stage as of February 2006. TinyOS 2.0 is devoted to maximize TinyOS platform portability and bear several enhancements from TinyOS 1.x, obtained through the systematic use of TEPs, TinyOS Enhancement Proposals (TEPs), for introducing and implementing new functionality in TinyOS.

---

[1] The official supported platforms are Linux RedHat 9.0, Windows 2000, and Windows XP.
[2] Source: http://www.tinyos.net/ – Over 10,000 downloads each new release, more than 500 research groups and companies employing the Berkeley/Crossbow Motes.
[3] Refer to: http://www.tinyos.net/scoop/special/working_group_tinyos_2-0.

1.2 Basic Concepts of TinyOS and neSC

Now that we have seen a bit of history and work in progress for TinyOS, we can look into some definitions and details, before exploring our WSN case study based on 2.4 GHz motes. Most of the descriptions followed are based on [TOS-DOC, GAY 03].

From the TinyOS mission statement[4] we see how they turned into implementation the basic ideas behind the NEST project. In short, TinyOS is an operating system designed to be run on devices with small memory footprint and maximum battery lifetime needs. It is intended for the development of event-driven applications, network communications and power management. It is highly adaptable, subject to reusability through its modular design featuring a component-based architecture. TinyOS component library includes network protocols, distributed services, sensor drivers, and data acquisition tools, all of which can be used as it is or be further refined for a custom application. It makes use of quickly executable inline code instead of using separate functions that would result in the need of intensive use of memory stacks.

A typical TinyOS application consists of a scheduler coordinating group of components, each one involved in response of certain events and command propagations.

Due to the nature of sensor networks, applications follow an event-driven architecture. Events could for example be generated in result of external changes, clock ticks or signal reception. Unlike traditional operating systems, employing kernel and memory manager concepts, hardware interactions here are through the components provided in the system library.

With reference to Figure 7.1, TinyOS provides a set of reusable system components, used by high-level components that may be bounded to them depending

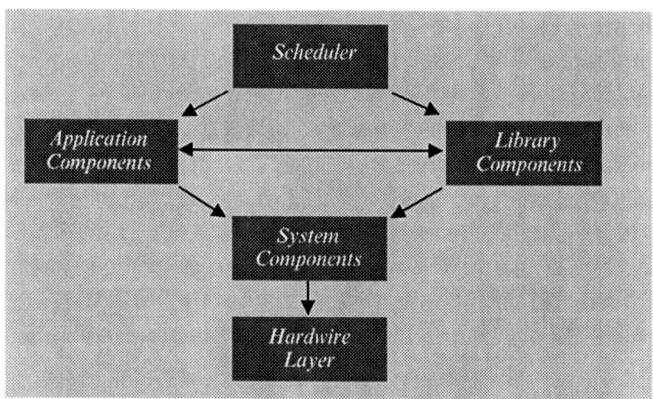

Fig. 7.1. TinyOS application architecture

---
[4] See: http://www.tinyos.net/special/mission.

on the application logic. Some components are real software modules, while others are wrappers around the hardware layers. This is transparent to developers, who do not need to care about the two types of components. The modular architecture makes TinyOS highly effective in that it enables the exclusion of unnecessary components, finally resulting in small memory requirements.

For instance, hardware-level components provide the implementation of certain services, like the transmission of a message, which can be used by high level components. High-level components in turn could provide call back functions when signalled by low-level components, like the reception of a message.

More formally, components communicate with each other through "commands" and "events". Commands propagate downward: they are issued by higher level components to lower level ones. Events propagate upward: they are signalled by lower level components and handled by higher level ones as shown in the Figure 7.2. The lowest level of events is generated by the hardware itself in the form of interrupts.

Other elements of the TinyOS component model depicted in Figure 7.2 are "frames" and "functions". A frame represents the internal state of a component, whereas functions are the implementation of commands and events described by interfaces.

Components provide and use interfaces for mutual interaction [TOS-DOC]. These interfaces are the only point of access to the component and are bidirectional. An interface declares a set of functions called "commands" that the interface provider must implement, and another set of functions called "events" that the interface user must implement. For a component to call the commands in an interface, it must implement the events of that interface. A single component may also use or provide multiple interfaces and multiple instances of the same interface.

Other key concepts in TinyOS are related to its memory usage and concurrency model. The first refer to the operating system design targeted to devices

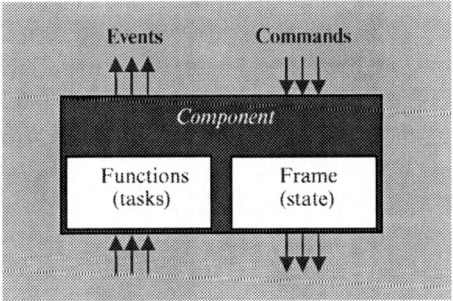

Fig. 7.2. TinyOS component model

with very limited memory. Components are loaded and linked together statically with the program size determined at compile time. It does not provide the heap for dynamic or runtime allocation of memory. It leaves the local variables on the stack: whenever flow is changed from one function to another, it temporarily saves the state to stack, using pointers to store variables.

As for the concurrency model, TinyOS executes only one program consisting of selected system components and custom components needed for a single application (see Figure 7.1), using two threads of execution: "tasks" and "hardware event handlers". Tasks are functions whose execution is deferred and that run to completion once scheduled. A task cannot be pre-empted by another task, it can rather be interrupted by a hardware event handler, which is executed and run to completion as well in response to a hardware interrupt.

Moving further in this brief description of the main elements of the NEST platform, we have not mentioned yet the programming language in which the TinyOS system is written: *nesC*, an extension to C designed to define and apply the structuring concepts and execution model of TinyOS.

The TinyOS Tutorial includes a quick overview of the nesC language concepts and syntax to get started with programming in this environment. In short, the TinyOS system, libraries, and applications are written in nesC, consequently there is a close relationship between the two. This language is primarily intended for embedded systems, such as sensor networks. nesC has a C-like syntax, but supports all the features required by TinyOS, such as the concurrency model and mechanisms for structuring, naming, and linking together software components into reliable network-embedded systems. Its high modularity by design allows developers to build components that can be easily composed into complete, concurrent systems. Moreover, it features extensive checking capabilities at compile time, looking for data races[5] and potential application deadlocks.

So, for readers who have some experience with C or C++ languages, writing simple nesC programs should be not too difficult, but it could be much more challenging developing complex applications that involve a large number of devices. There needs to fully understand the models and paradigms behind TinyOS and nesC, particularly configurations, which link together components.

One of the main difficulties with TinyOS and nesC is to have access to clear and up-to-date information. The tutorial we already mentioned is a little outdated (2003), but for the reader interested in a good comprehensive introduction to TinyOS programming, there is a recent guide (February 2006) available on-line [LEV 06].

We will see practical examples, operational procedures and sample TinyOS applications in the upcoming sections.

---

[5]nesC programs are susceptible to certain race conditions in accessing data, because "tasks" and "hardware event handlers" may be pre-empted by other asynchronous code.

1.3 Case Study: Crossbow Motes

The overall scope of this case study is to show a sample set-up of a small WSN by using current technology widely available. Our experiments are based on the Crossbow's Mica-Z platform, a member of the Mica mote family, operating in the 2.4 GHz ISM band. Some quick definitions follow.

A "mote" is a processor/radio board which includes a central processing unit (CPU), memory, an analogue/digital (A/D) converter for reading sensor data and a radio transceiver. When coupled with a sensor or data acquisition board, it becomes a full WSN node.

There is a commercially available product, called "mica mote" that has been used widely by researchers and developers. It has all of the typical features of a mote and is available to the general public through a company called Crossbow, under the original design of UC Berkeley. Figure 7.3 shows a Mica-Z mote (top view) compared to a 1.5 V AA battery. Two of them are necessary to power it for more than a year, depending on the particular running application and duty cycle, as well as nominal battery capacity chosen (typical 2000 mA/h). On the bottom side, not visible in figure, there is the attached battery pack hosting a pair of cylindrical AA size batteries. More information can be found in Crossbow User's Manual [XBO 05b].

To programme a mote, we need an interface board, typically multipurpose, used in conjunction with a specific mote family of products. It provides communication and in-system programming. Sometimes this programming board also features a "Power over Ethernet" connection, available for the Crossbow's MIB600 model we used for the experiments.

To conclude our brief introduction on mote hardware, Crossbow Technology, Inc. based in San Jose, California is the manufacturer of a wide range of products for WSN besides Mica-Z motes; on their catalogue we can also find different kits to test and deploy complete solutions for various applications (e.g. building

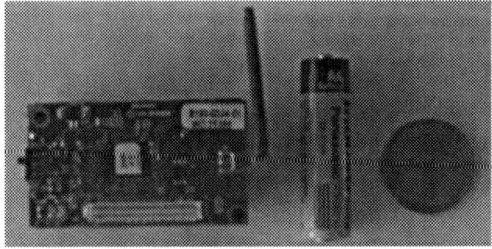

Fig. 7.3. Crossbow Mica-Z mote (W 58 × L 32 × D 7 mm)[6]

---

[6]Excluding battery pack.

automation, navigation, environmental monitoring). They are named on the WEBS Research Centre page of UC Berkeley under project collaborators (as of the date, they are the sole explicitly cited).

Objectives and Test Cases. The main objectives of the experiments are to implement a field trial for WSNs and at the same time verify the operation of a sample application capable of multihop routing.

Related to those, other interesting objectives are:

- To make some performance measurements and preliminary tests
- To figure out a repeatable set of experiments
- To get a basic characterization of mote's communication

As we will see along the description, the following test cases are analysed:

- Indoor simple environment (single office)
- Indoor long hallway
- Indoor premises (multiple areas of a building floor)

Action Plan and Activities. To perform the job some decisions and practical steps are to be made:

1. Choice of one or more initial indoor environments (this a supposed target for WSN, examined in other testbeds, see [ZHA 03])
2. Preparation of plans for tests and experiments
3. Integration and/or adaptation of several pieces of TinyOS code as needed
4. Installation, tuning and configuration of all the necessary equipment (hardware and software)
5. Collection and analysis of a set of repeatable measures

The result of these activities will be evident next in the chapter.

Tools, Softwares and Lab Set-up. Figure 7.4 shows our basic network lab set-up, used for actual operation and data collection/analysis as well.

The following list contains the available equipment we made use of at the time of experiments:

- Remote motes: up to 4 Mica-Z motes (2.4 GHz), programmable RF output power (from −25 dBm to 0 dBm), mote's group_id (default 125),

Practising

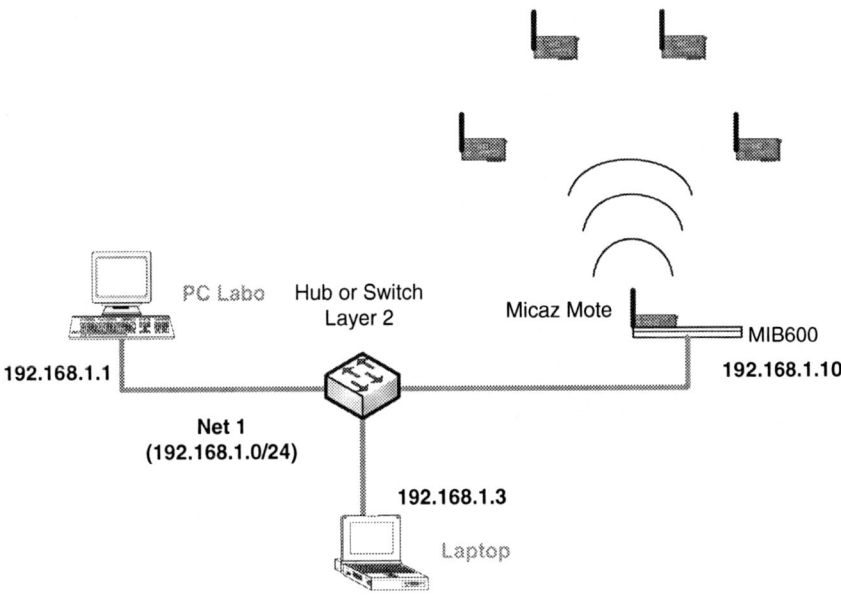

Fig. 7.4. Lab basic network diagram

frequency channel (from chapter 11 to chapter 26 in the IEEE 802.15.4-allocated band, 5 MHz spacing)

- Base station: MIB600 programming board (TCP/IP communication to the programming and controlling environment), 1 mote programmed as a base station

- PC side: Java tools and executables developed for collecting and analysing TinyOS data (e.g. SerialForwarder, Surge-GUI and related logging and statistical tools, Xlisten, etc.) mainly packetized by Crossbow;

Note: The internal transceiver of Mica-Z motes (Chipcon CC2420) is IEEE 802.15.4 compliant and could be used to build ZigBee networks. However, it must be noticed that the current MAC-layer in TinyOS does not implement an IEEE 802.15.4 medium access, that is under development at present.[7]

Other details of our set-up are in Table 7.1 below.

Table 7.2 summarizes Crossbow's Mica mote product family and Moteiv's Telos platform, features and capabilities (adapted from [POL 05] and [XBO 05b]).

Notice that for our basic experiments we are mainly interested in layers 1 and 2 network behaviour, consequently we do not cover sensor reading (and

---

[7] See the (Tinyos-devel) list and the TinyOS FAQ, both available at: http://www.tinyos.net.

Table 7.1. Lab Set-up details

Equipment	Description/details	Qty.	Notes
Control and programming station	Standard PC	1	WinXPpro SP2, Intel Pentium II – 930 MHz, 256 MB RAM
Programming environment	TinyOS 1.1.10 + Cygwin	1	Crossbow's fully tested to date
Network	Ethernet hub 8-port	1	Suitable also for network sniffing
Wireless network (optional)	AP/bridge and/or client-bridge	2	To get optional remote connection to the base station (MIB600+ mote id-0)
Java tools	Surge (GUI), Stats, SerialForwarder, HistoryViewer	1	Last version packetized by Crossbow (Surge-View zip)
Other tools	Xlisten	1	Ver.1.2
Laptop PC (optional support)	Standard notebook	1	Pentium-M, WinXPpro SP2, network tools
Programming board	Ethernet programming board	1	Crossbow part-number MIB600CA
Network nodes	Mica-Z motes (2.4 GHz)	5	Crossbow part-number MPR2400CA
Sensor boards (optional)	"Micasb" sensor boards (temperature/light/audio)	2	external sensors – Crossbow part-number MTS300

applications) aspects in detail. Moreover sensor boards are energy consuming, so it is advisable not to attach them if not necessary.

Preparing the Environment. Installing the TinyOS environment on a PC is "relatively" simple following directions on the project web site (http://www.tinyos.net/tinyos-1.x/doc/install.html). Assuming a Windows box as in our lab set-up and starting from scratch, it is advisable installing with the "InstallShield wizard" rather than manually. This includes all the packages and tools needed to get to a certain baseline version (i.e. 1.1.0).

The basic TinyOS 1.1.0 InstallShield set-up offers the following software packages:

- TinyOS 1.1.0

Table 7.2. The family of Berkeley notes and their capabilities

Mote type	Mica	Mica2Dot	Mica 2	Mica-Z	Telos
Year	2001	2002	2002	2004	2004
Microcontroller					
Type	Atmega103L (4 MHz)	Atmega128L (4 MHz)	Atmega128L (7 MHz)	Atmega128L (7 MHz)	TI MSP430
Program memory (KB)	128	128	128	128	48
RAM (KB)	4	4	4	4	10
Active power (mW)	8	8	33	33	3
Sleep Power (µW)	75	75	75	75	15
Wakeup time (µs)	180	180	180	180	6
Nonvolatile storage					
Chip	AT45DB041B	AT45DB041B	AT45DB041B	AT45DB041B	ST M25P80
Connection type	SPI	SPI	SPI	SPI	SPI
Size (KB)	512	512	512	512	1024
Communication					
Radio	TR1000	CC1000	CC1000	CC2420	CC2420
Frequency band (MHz)	433/915	315/433/915	315/433/915	2400	2400
Data rate (kbps)	40	38.4	38.4	250	250
Modulation type	ASK	FSK	FSK	O-QPSK	O-QPSK
Receive power (mW)	12	29	29	38	38
Transmit power at 0 dBm (mW)	36	42	42	35	35
Power Consumption					
Minimum Operation (V)	2.7	2.7	2.7	2.7	1.8
Total active power (mW)	27	44	89	71	41
Programming and sensor interface					
Expansion	51-pin	19-pin	51-pin	51-pin	16-pin
Communication	IEEE 1284 (programming) and RS232 (requires additional hw)				USB
Integrated sensors	No	Yes	No	No	Yes

- TinyOS Tools 1.1.0
- NesC 1.1.0
- Cygwin (package trees from approximately April 2003)
- Support tools
- Java 1.4 JDK and Java COMM 2.0
- Graphviz
- AVR Tools

After that, the easiest way to upgrade is via RPMs, a Linux-style packet manager, with available concurrent versions system (CVS)[8] snapshots. RPMs can be used on both Linux and Cygwin/Windows, the latter being a Linux-like environment for Windows (a kind of emulation).

A CVS snapshot release is a convenient-to-install package, but it must be noticed that it could be not fully tested by anyone other than the implementor.

For example, the TinyOS 1.1.10 snapshot is the preferred Crossbow's version as reported on [XBO 05a], a guide that we suggest to use as an alternative to the link above if working with a Crossbow's kit as we did.

Working with Crossbow's applications (under/opt/tinyos-1.x/contrib/xbow/apps), it is possible to set most of the parameters by editing the "MakeXbowlocal" file (/opt/tinyos-1.x/contrib/xbow/apps/MakeXbowlocal).

For setting the RF power level ("MICAZ RF Power Levels" section of the file), uncomment the corresponding line, such as:

$$CFLAGS+ = -DCC2420\_TXPOWER = TXPOWER\_M3DBM$$

Similarly, to select a frequency channel for Mica-Z motes, there is a section named "ZigBee Channel Selection", where it is possible to pick up the desired operating frequency, for instance:

$$CFLAGS+ = -DCC2420\_DEF\_CHANNEL = 26$$

All the things work together at compile time thanks to a particular "Makefile" that each application must have in its root directory (e.g. /opt/tinyos-1.x/contrib/xbow/apps/Surge_Reliable/ for the application Surge_Reliable).

Notice that the application "Makefile" has to take into account the "MakeXbowLocal" file and the latest make system, adding these lines:

include ../MakeXbowlocal

include ${TOSROOT}/tools/make/Makerules

Then, if the software environment is correctly set, to compile a TinyOS application it is necessary to issue a "make" command from a Cygwin shell in the program directory, such as:

$ make micaz

This "make" compiles the application for the target platform, here micaz, and creates a sub-directory tree as /build/<platform_name> (i.e. "build/micaz/"), containing the executable for the specified hardware type.

---

[8] It is a widely used revision control system for software development, popular in the open-source world.

Finally, the program can be actually uploaded to a target mote as showed below (refer to the lab set-up in Figure 7.4):

$ make micaz reinstall,75 eprb,192.168.1.10

This command uploads a previously compiled application with the parameters:

- \<micaz\>   platform type (e.g. mica, mica2, micaz)
- \<reinstall\>   upload an executable already compiled for a target[9]
- \< 75 \>   node-ID (mote's local address to be set)
- \<eprb\>   the type of programmer in use (i.e. Ethernet Programming Board (eprb), MIB600)
- \<192.168.1.10\> IP address configured on the programming board

The mote is now programmed and the same steps apply to preparing a group of motes sharing an identical application.

In order to listen to the information coming from the WSN, it is necessary to run a socket listener on a PC, like the Java-tool named *SerialForwarder*. "SerialForwarder" is a Java application that acts as a bridge between different connections. It provides TCP/IP communication at the back-end side for one or more controlling or monitoring clients.

It is used to read packet data from a computer's serial port or network link and forward it over a server port connection, so that other programs can communicate with the sensor network via a sensor network gateway. "SerialForwarder" does not display the packet data itself, but rather updates the packet counters in the lower-right hand corner of its control window. Once running, the Serial Forwarder listens for network client connections on a given TCP port (9001 is the default), and simply forwards TinyOS messages from the communication port (serial or network) to the network client connection, and vice versa. It is important to notice that multiple applications can connect to the Serial Forwarder at once, and all of them will receive a copy of the messages from the sensor network.

In Figure 7.5 we can see typical configuration parameters of SerialForwarder, that is:

- \<Server Port\>   9001 (default)
- \<Mote communications\> network@192.168.1.10:10002

---

[9]This improves operational speed when there are several motes to install.

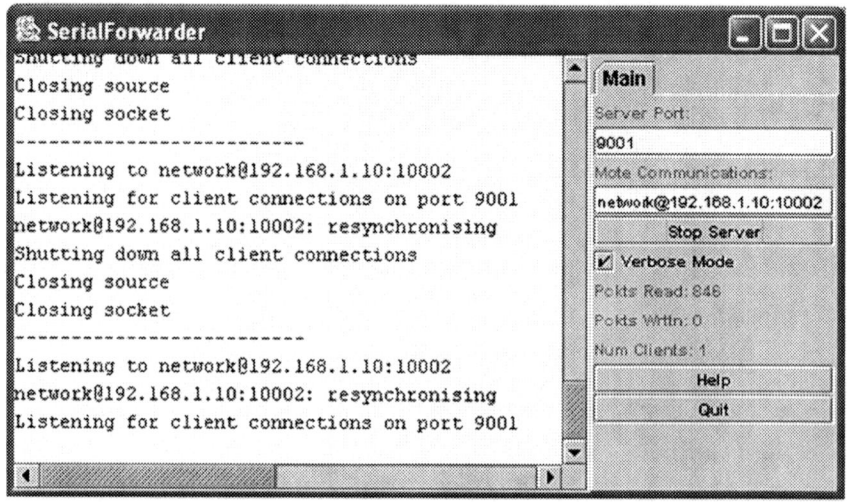

Fig. 7.5. SerialForwarder control window

The general syntax for mote communications, depending on how the host PC is attaching to the base station, is:

serial@COM<#>:57600

where <#> = 1, 2, 3, etc., for serial port COM1, COM2, COM3, etc., if using a serial interface board (e.g. Crossbow's MIB510), or:

network@<IP_Address_of_MIB600>:10002

which contains the IP address of the ethernet board MIB600 and the default TCP port, the latter corresponding to the serial interface of the attached mote (i.e. mote ID-0) through the virtual serial server of the board itself.

Our WSN is now connected to a classical IP network. Then, to write programs capable to read data from the WSN, there is a very powerful tool: MIG.

MIG (message interface generator) is a tool that is used to automatically generate Java classes that correspond to active message types used in mote applications. MIG reads in the nesC struct definitions for message types in a mote application and generates a Java class for each message type that takes care of the tricky details of packing and unpacking fields in the message's byte format.

MIG is used in conjunction with the net.tinyos.message package, which provides a number of routines for sending and receiving messages through the

MIG-generated message classes. For example, nesC constant generator (NGC) is a tool to extract constants from nesC files for use with other applications and is typically used with MIG.[10]

1.4 Experimentation and Measurements

Multihop in a Simple Enclosed Environment. This experiment is designed to test and integrate different components already available in the TinyOS world for multihop communication in a WSN, by an application called Surge. In subsequent sections we will also see snapshots and outputs from some of the softwares we used for data collection and analysis (Surge-View tools), such as Surge-GUI, Stats, etc., freely downloadable at Crossbow's support site (http://www.xbow.com/support/).

Test Preparation and Layout. For this first experiment we made use of the same laboratory room employed for system check and validation of the Crossbow kit. As showed in the figure below, this is a real office, $\sim 6 \times 7$ m, fitted with chairs, tables, PCs monitors, etc.

We wanted to test basic multihop routing with Surge_Reliable, an application provided by Crossbow for Mica-mote WSNs (see /tinyos-1.x/doc for details about Surge). In order to simulate a kind of simple radio frequency (RF) environment for radio propagation between sensor nodes, we regulated the output power of the Mica-Z motes at a very low level (-15 dBm or -25 dBm), so that each closer pair would see a direct optimal path. This can be seen also as a simulation of farther distances at higher power levels.

Furthermore we paid attention in placing the motes in an imaginary horizontal plane, above the level of possibly interfering objects in the office (>1 m). Similarly, the base station and the other motes were put farther the area where the PC control station and the tester were operating.

Because of the functioning of the Surge_reliable algorithm (two-way communication), when adjusting power levels we also cared to keep things symmetric, that is all the motes with the same RF output to not incur in odd behaviours.

Methodology and Activities. The area where we made our experiments is generally a floor covered with linoleum tiles 30×30 cm, walls are either in concrete/bricks or pre-built plastic/metal structural panels, doors are in wood or metal, ceilings have the typical structure for interior premises (suitable to accommodate lights, electrical cables, etc.). Hence we could get distance measurements with a ruler, as well as counting the floor tiles.

---

[10]Extract from the TinyOS Tutorial.

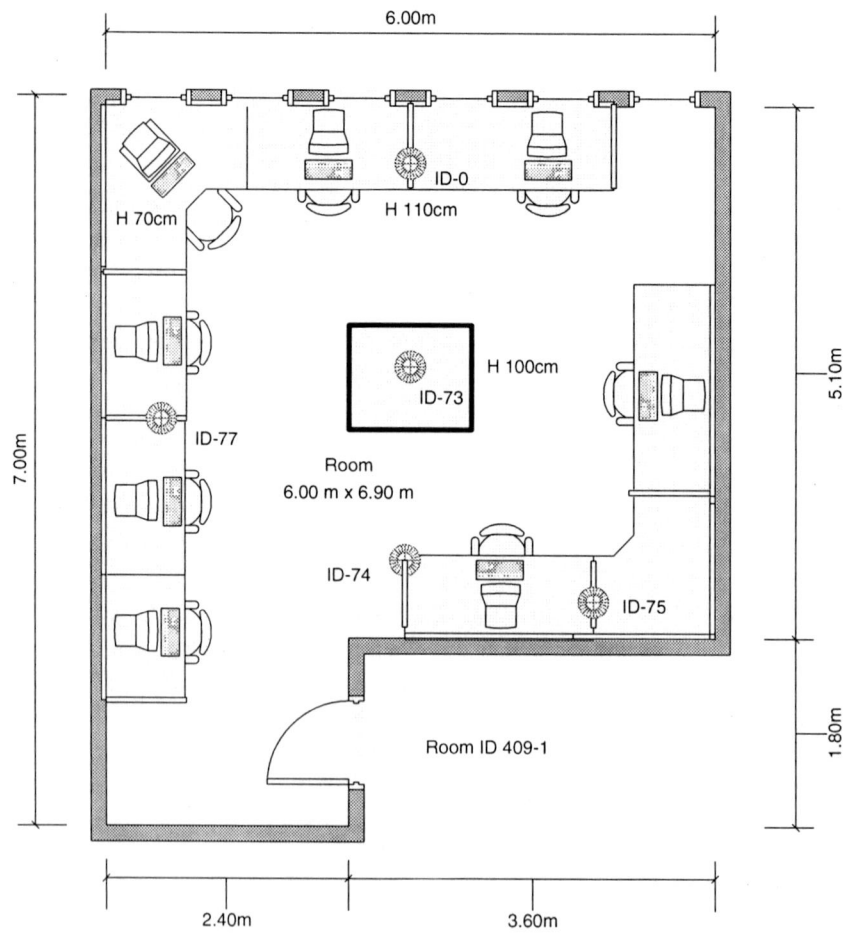

Fig. 7.6. Lab room with motes

The overall placement of the motes in our office environment is displayed in Figure 7.6 with a circular point as well. Then, starting from the device closer to the base (ID-0), each node was turned on one at a time to see what happened and how network would form over time, in the following order:

(a) Mote-id 73

(b) Mote-id 74

(c) Mote-id 75

(d) Mote-id 77

Practising 237

After complete set-up, a fault on a critical node was simulated (shutting it down) and network reaction was observed. A log file during every step was collected for statistical analyses and post-processing.

Experiment Set-up. Table 7.3 shows the specific parameters chosen for the experiment.

Tools at the PC Side. After installing all the control programs packetized by Crossbow under "C:/Program Files/Surge-View" of a Windows host PC, the following steps are required:

(a) Run the socket listener for TinyOS packets (SerialForwarder.exe)

(b) Configure the SerialForwarder server to connect to a specific interface board (i.e. MIB600)

(c) Start the Surge-GUI application provided (Surge.exe) with the correct parameters

(d) (optionally) Redirect the standard output of "Surge.exe" to a log file for post-processing

Table 7.3. Set-up for the multihop experiment

Parameter	Value	Notes
MOTE GROUP_ID	$0 \times 7D$	Decimal 125, default
MOTE NODE_ID (base station)	ID 0	ID 0 (mandatory) for the mote base station placed on top of the programming board MIB600;
MOTE NODE_IDs (others)	IDs 73, 74, 75, 77	ID equals to the last two digit of the Crossbow mote's serial number for the remote motes;
MOTE FREQUENCY CHANNEL	IEEE 802.15.4, chapter 26 (2480 MHz)	It is supposed not to overlap with 802.11 channels
MOTE RF OUTPUT POWER	-15 dBm and -25 dBm	Adjusted via the MakeXbowlocal file
SURGE TIMER RATE	8	One packet every 8 s, suggested initial rate[11]

---

[11] Parameter INITIAL_TIMER_RATE, file Surge.h under "/opt/tinyos1.x/contrib/xbow/apps/Surge_Reliable/".

The Surge executable is capable to connect to SerialForwarder on port 9001, collect and display TinyOS Surge packets coming from the WSN through the base station (mote ID-0 together with ethernet interface board MIB600).

The box below shows a typical session at the PC side (commands in bold).

```
C:\Program Files\Surge-View>Surge 125> surge-1n-260905.log

C:\Program Files\Surge-View>Stats<surge-1n-260905.log
Node Number: Packets Received: Packets Sent: Success Rate: Parent Changes:
  Level Changes: Average Level: Duty Cycle: Battery Voltage:
0: 35: 35: 1.0: 1: 0: 0.0: 0.0: 3.270460358056266
75: 38: 38: 1.0: 0: 1: 0.9210526315789473: 0.0:2.987733644859813

C:\Program Files\Surge-View>Surge 125

C:\Program Files\Surge-View>node Number#Message Count#String Date#
Time#interval#parent#Message Rate#Sequence Number#hopcount#mAm#
Batt#id 0#hopount 0#quality 0#id 1#hopount 1#quality 1#id 2#hopount 2#
quality 2#id 3#hopount 3#quality 3#id 4#hopount 4#quality 4#Temp#Light#
```

Other available tools are "Stats.exe" and "HistoryViewer.exe", which give an overview of the health of the Surge sensor network.

Practical Results – Multihop Behaviour. At first, with motes at their full default output power level (0 dBm), it was not possible to get a multihop behaviour. Then power was decreased until a suitable value for this environment was reached between -15 and -25 dBm with an average mote's distance of 2 m.

As for the Mica-Z datasheet, the maximum range is:

- 75 m to 100 m outdoor (line of sight), half-wave dipole antenna

- 20 m to 30 m indoor, half-wave dipole antenna

However, the antennas which normally come with Mica-Z motes are quarter-wave whip types (monopole), which are less effective[12].

Snapshots of the network were taken with motes at -25 dBm output power, showing how network was forming as the number of motes increased. Nodes in the topology window were identified and placed in their actual relative position (see Figures 7.6 and 7.7).

---

[12] On the FAQ they say to expect about half that distance without a ground plane that is usually the case for quick deployments.

Practising 239

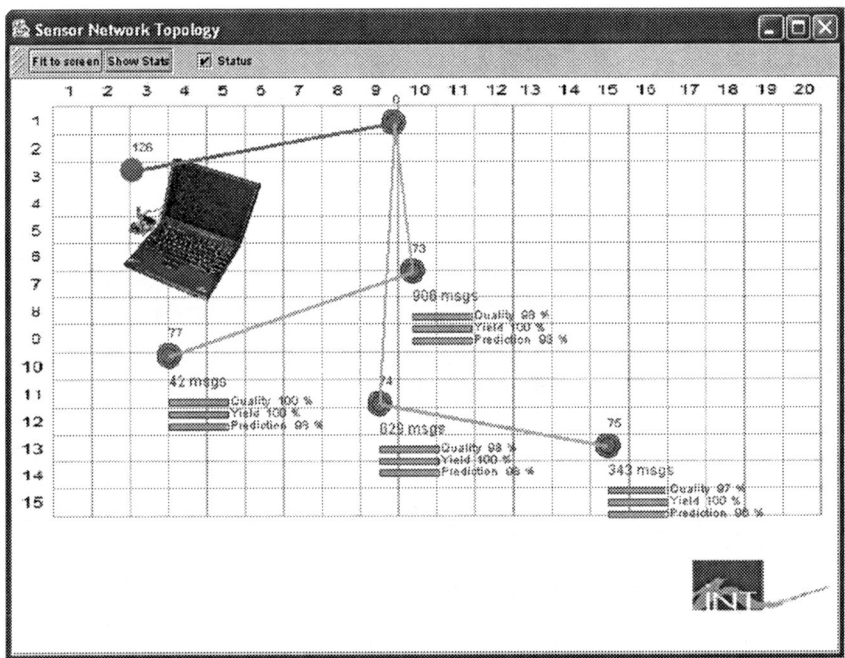

Fig. 7.7. Example sensor network topology with Surge tools

Next to each node are also reported some real-time statistics, that is:

- Received messages at the base from node
- Quality (an internal parameter for the Surge algorithm)
- Yield (a measure of successful reception, that is: number of packets received at base/number of packets sent from node)
- Prediction[13]

For the time of the experiments the overall statistics for packet delivery was almost completely successful (however here we were just looking at network formation for a relatively short period of time).

An extract from Surge output files, adapted for readability, is reported in Table 7.4.

Node 0 is the mote base station, collecting all the packets from the WSNs. This mote itself generates Surge packets delivered, via the MIB600 programming board, to the PC control station running Surge-GUI and SerialForwarder over a TCP/IP connection.

---

[13] A weighted estimation of future performance based on current success rate (Yield field) and quality.

Table 7.4. Surge_reliable statistics (no faults on the network)

Node	PktsRec	PktsSent	SuccRate	Parent changes	Level changes	AvLevel	Duty cycle	Batt
0	897	897	1.0	1	0	0.0	0.0	3.27
73	1134	1134	1.0	0	1	0.987	0.0	2.82
74	861	861	1.0	0	1	0.983	0.0	2.84
75	579	580	0.998	2	2	1.642	0.0	2.86
77	282	282	1.0	1	1	1.900	0.0	2.98

Each node, base station mote-0 included, accounts for itself solely, even though it could pass other packets over for routing. Node 73, 74, 75 and 77 are remote motes. Notice that farther nodes have a "Level" in the network hierarchy greater than one, which is a symptom of multihop routing to the base.

An example of a full Surge_reliable raw packet, as logged on the PC control station, is showed below (node number and parent fields highlighted):

---

node Number#Message Count#String Date#Time#interval#**parent**# Message Rate#Sequence Number#hopcount#mAm#Ba|t#id 0#hopount 0#quality 0#id 1#hopount 1#quality 1#id 2#hopount 2#quality 2#id 3# hopount 3#quality 3#id 4#hopount 4#quality 4#Temp#Light#

0#895#10/11/2005 02:30:54 PM#1129033854057#2002#**126**#0.5000350024501715#911#0#0#391#0#0#0.0#0#0#0.0#0#0#0.0#0#0#0.0#0#0#0.0#102#85#52#100#255#255#

75#582#10/11/2005 02:30:54 PM#1129033854178#1432#**0**#0.707093562620206#631#1#0#447#0#0#0.878431372549 0196# 74#1#0.984313725490196#73#1#0.6745098039215687#77#2#0.0#0#0#0.0#101#87#70#98#0#0#

74#869#10/11/2005 02:30:54 PM#1129033854428#1422#**0**#0.7035119315623592#912#1#0#449#0#0#0.98431 3725490196#73#1#0.996078431372549#75#1#0.984313725490196#77#2#0.0#0#0#0.0#102#88#75#102#0#0#

73#1140#10/11/2005 02:30:54 PM#1129033854458#1422#**0**#0.699467006141 203#1186#1#0#452#0#0#0.9921568627450981#74#1#0.9921 568627450981#77#2#0.9529411764705 882# 75#1#0.9372549019607843#0#0#0.0#103#89#66#100#0#0#

77#288#10/11/2005 02:30:54 PM#1129033854708#1392#**73**#0.7151029748283753#336#2#0#430#73#1#0.921 5686274509803#0#0#0.6470588235294118#75#2#0.0#74#1#0.0#0#0#0.0#100#84#65#100#0#0#

Practising 241

It can be noticed the rather high number of parameters used to make the multihop algorithm work, either for accountability or for operation.

The meaning of some of the fields of the log file above is as follows:

- **Node number** Node identifier
- **Message count** Number of messages received (at base)
- **Sequence number** Packet ID as sent from node
- **Hopcount** Number of hops between base station and node
- **Parent** Current routing parent (highlighted)
- **Quality** Link quality to node-i (i.e. for neighbour 0, 1, 2, 3, 4)
- **Voltage** Battery voltage[14]

Another part of the experiment was to simulate a fault condition, choosing a critical node like mote ID-73 and turning it off. At the beginning (first initial minutes) the counters on the impacted nodes stop (mote 73 and 77). After a while the still active node (mote 77) tries to re-enter the network by node 74, and then its internal algorithm chooses to connect directly to the base station. Obviously the overall statistics for node 77 go down because many packets were lost during the outage. If letting it run for a long time in good network condition, the relative success rate value (Yield field) will recover to higher values as usual. Node 73 appears alone on the topology window and after few minutes goes away (not showed).

Experiments with other topologies and power levels were carried out to look at different cases. An example of Surge statistics is given along, together with some explanation.

In Figure 7.8 some interesting pieces of information can be noticed:

- *Success rate values (Yield field in Surge terminology) greater than one* are symptoms of signal multipath, since in this basic test we are just collecting packets at base without further processing; during data post-processing it needs to be careful, filtering unwanted duplicates

- *P1 and P2 indicate alternative paths* to base station in the form of most common parents; the Surge algorithm uses its best candidate, ready to switch to another path if necessary

The "Stats" application we mentioned at the beginning among available PC tools, deals with multiple reception of the same packets correctly. This was

---

[14] In ADC (Analog to Digital Converter) count, to be converted in engineering units.

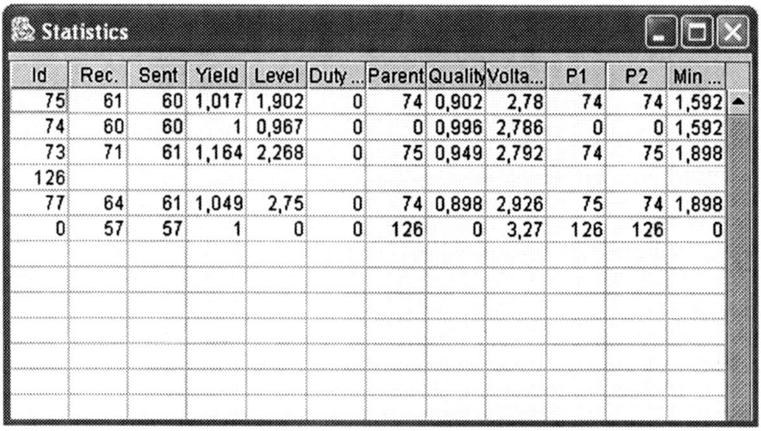

Id	Rec.	Sent	Yield	Level	Duty...	Parent	Quality	Volta...	P1	P2	Min...
75	61	60	1,017	1,902	0	74	0,902	2,78	74	74	1,592
74	60	60	1	0,967	0	0	0,996	2,786	0	0	1,592
73	71	61	1,164	2,268	0	75	0,949	2,792	74	75	1,898
126											
77	64	61	1,049	2,75	0	74	0,898	2,926	75	74	1,898
0	57	57	1	0	0	126	0	3,27	126	126	0

Fig. 7.8. Example statistics window with Surge tools

tested by processing a small log file of Surge data with a spreadsheet program and comparing the results.

Another interesting tool for capturing raw packets from a WSN is "**Xlisten**"[15], which is a standard application that can be found in "contrib/xbow/apps/" (see also the TinyOS Tutorial for similar tools). It has several options, as can be seen in the Crossbow's Getting Started Guide, some of them are:

– < −r > Tell Xlisten to display raw data coming in from sensor network

– < −p > Tell Xlisten to try to parse bytes from the actual data message

– < −c > Tell Xlisten to try to "cook" bytes (convert it in engineering units) from the actual data message

Since Xlisten has been mainly designed for Surge packets, the options "−p" and "−c" are only useful for known packet types (source code is available for modifications in the same own directory). Nonetheless, when used with the option "−r", it is a powerful tool to see raw packets coming in from a sensor network, for example placing a mote with the TOSBase application as a gateway that simply forwards packets to the serial interface and then back to a host PC.

This is in fact what we did to capture messages for applications like "Cnt-ToLedsAndRfm" and "XSensorMTS300" (see [XBO 05a]). From the Xlisten output can be easily noticed that those TinyOS applications generate broadcast messages to the address FFFF.

---

[15] Under "contrib/xbow/tools/src/xlisten/".

Packet Reception Along an Indoor Hallway. This experiment is designed to get some basic performance data for mote's communication, using and integrating different applications for packet transmission/reception. We will show some preliminary results for packet success and packet loss rate in a testbed composed of only two motes (direct communication link). An effort was made to validate outcomes obtained with certain tools, trying other methods in the same environment and context. We started with the same "Surge" application we examined in the previous experiments, and then moved on using and customizing another couple of TinyOS programs, "CntToLedsAndRfm" and "Pong", as we will discuss in detail in the following section.

Test Preparation and Layout. For this experiment we looked around for a suitable indoor environment and found a long hallway just on the floor where we had our lab room. As shown in Figure 7.9, this is a building with a horizontal side more than 30 m long.

We wanted to make some measurements of layer 1 communication between a couple of motes getting inspiration from work in [ZHA 03] (where packet delivery performance is studied with up to 60 Mica motes placed in different environments, indoor and outdoor) and carry out preliminary tests in a quick and straightforward way.

In order to correlate samples taken on the corridor and distance from base station, we put marks along it with a granularity of 1.5 m, using small adhesive strips every five floor tiles. Furthermore we paid attention in placing the motes right in the passage centre for symmetry, as well as laying them on small cardboard boxes, to separate them a little from the floor and get a less harsh propagation environment.

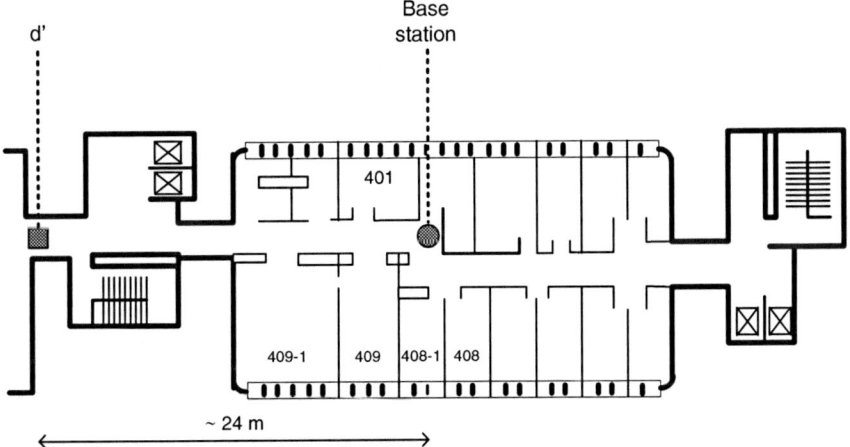

Fig. 7.9. Plan of the building floor used for experiments

Note: During the tests there were sometimes in progress some experiments with IEEE 802.11b equipment in an office nearby (access points on Linux), working in the same 2.4 GHz band. When capturing traffic with an air-sniffer, they occasionally caught errors on frame, supposing some form of interference with motes. However, at the beginning our sensor network operating on ZigBee channel 26 (2480 MHz) seemed not to suffer from IEEE 802.11b, as suggested in Crossbow's technical paper on avoiding interference between Wi-Fi and ZigBee (see [XBO TR]).

Methodology and Activities. As we already saw in the previous section, the area where we made the experiments is a floor covered with linoleum tiles 30×30 cm, walls are either in concrete/bricks or pre-built plastic/metal structural panels, doors are in wood or metal, ceilings have the typical structure for interior premises (suitable to accommodate lights, electrical cables, etc.).

Since we could not cover the whole distance with available motes, we used just a couple of motes, one being the base station, the other the remote mote, to gather measurements along the corridor. Then, starting from the first mark we placed during test preparation (that is 1.5 m from base station), we moved on step by step, waiting to collect a certain number of samples before proceeding further to another point.

The experiment was repeated at different days/times and with more than one tool/application to get a sufficient set of measures and double-check results. Several log files were collected for statistical analyses and post-processing activities.

Experiment Set-up for Surge. Since we were interested in packet reception behaviour over distance, at first we tried to reuse the same tools as we did for tests inside the lab, as everything was already available and configured to collect sequenced packet transmitted from a remote mote to a base station.

The following parameters (Table 7.5) were chosen for the experiment, similarly to what we did for multihop with Surge previously, but adapted for this specific context and then with higher power levels.

Experiment Set-up for Pong. To sample received signal strength indicator (RSSI)[16] and link quality indicator (LQI)[17] values at mote's receiver, it is possible to use "Pong", an application developed for Telos, a wireless sensor device like Mica-Z manufactured by MoteIV (www.moteiv.com).

We then managed to modify a little the suite of software components to make them work with our motes and environment (TinyOS-1.1.10). Compile and

---

[16] Received Signal Strength Indicator.
[17] Link Quality Indicator.

Table 7.5. Set-up for the hallway experiment

Parameter	Value	Notes
MOTE GROUP_ID	0 ×7 D	Decimal 125, default
MOTE NODE_ID (base station)	ID 0	ID 0 (mandatory) for the mote base station placed on top of the programming board MIB600;
MOTE NODE_IDs (other)	IDs 74	ID equals to the last two digit of the Crossbow mote's serial number for the remote motes;
MOTE FREQUENCY CHANNEL	IEEE 802.15.4, chapter 26 (2480 MHz)	It is supposed not to overlap with IEEE 802.11 channels
MOTE RF OUTPUT POWER	−3 dBm and 0 dBm	Adjusted via the MakeXbowlocal file
SURGE TIMER RATE	8	One packet every two seconds, maximum rate suggested for Surge

install on two Mica-Z motes, one with ID 1 (mandatory) and another whatever (ex. 77); then the one with ID 1 has to be placed on MIB600 as base station and the other far away, both turned on (the first on the base gets energy from the programmer itself, do not turn the power switch on for that).

The following specific parameters were used to compile the corresponding ".nc" code:

- Mote Rf Output Power: −3 dBm

- Counter Timer Rate: 2000 or 200 (one packet every 2 s or one packet every 200 ms)[18]

To just see raw packets, it is possible to use "xlisten-r" (see the previous section) in order to get data coming from the mote attached to base station and have a look at those by hand (to decode it, see also the TinyOS Tutorial lesson 6); when reaching the payload field, use "PongMsg.h" under the application directory to interpret the data.

---

[18] Timer interface in "/opt/tinyos-1.x/apps/Pong/PongM.nc" file.

A smarter way to decode packets is using the java tools coming with Pong, i.e. "PingPong.java". To compile it, issue the following command in a Cygwin shell:

$ javac PingPong.java

After that, just run "SerialForwarder" as seen before and launch:

$ java PingPong

to see RSSI values and LQI values in Pong messages, brought to us by SerialForwarder talking to our MIB600.

Actual values embedded in Pong packets are easily seen with the related java tools, as shown in the box below.

```
$ java PingPong
1,−13,−58,107,High,77,−13,−58,107,High
```

Let's have a look at the line in PingPong output above. The meaning of the fields is as follows:

(a) First half, values referred to mote ID-1 at base station:

Source-ID	1
Source_RSSI (at transceiver)	−13
Source_RSSI (in dBm)	−58
Source_LQI	107
Source_LQI (interpreted)	High

(b) Second half, values referred to remote mote ID-77:

Destination-ID	77
Destination_RSSI (at transceiver)	−10
Destination_RSSI (in dBm)	−55
Destination_LQI	107
Destination_LQI (interpreted)	High

RSSI in dBm is obtained by subtracting a fixed number (i.e. 45, corresponding to the approximate RF front-end gain) from values detected at transceiver (see Chipcon CC2420 datasheet).

Similarly, LQI and its range are interpreted from values reported in the same datasheet as follows:

– LQI $>= 104$ is "high"

– LQI between 80 and 103 is "medium"

– LQI < 80 is "low"

Practising 247

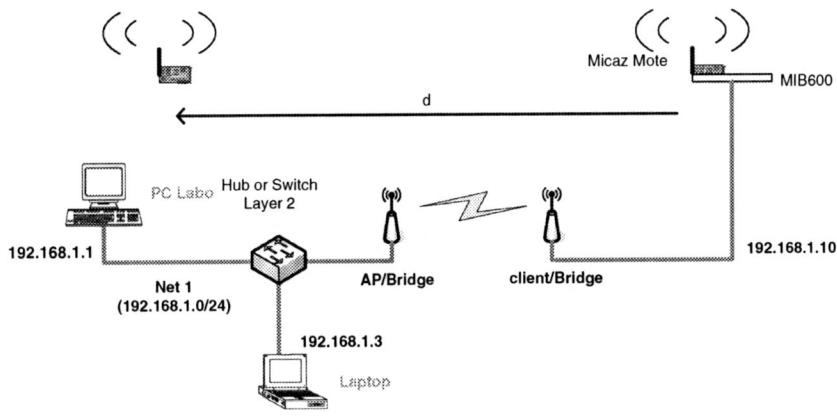

Fig. 7.10. Lab network diagram with optional remote link

Tools at the PC Side and Network Set-up. Software tools and equipment in the back end mentioned until now, as well as practical information, have been already given in previous sections and will not be repeated here. A graphical illustration of the overall network setup and testbed for experiments over distance discussed so far is shown in Figure 7.10.

An optional remote link with Wi-Fi equipment could be useful to repeat tests in outdoor or in remote areas well far away from the control stations. For these experiments, since we ran several test sessions, we prepared and made use of ad-hoc forms to record set-up parameters and log important data.

Practical Results – Packet Reception. Papers on WSNs and available resources on TinyOS lead to the feeling that performance is strongly related to the expected applications and typical contexts, which translates to systems designed for packet delivery of short messages (e.g. sensor readings) at a slow pace for a long time (e.g. [ZHA 03]).

That is why many presentations and reports focuses, depending on the situation and environment, on packet loss or packet success rate, maximum packet rate vs number of motes (when studying layer 2 communication), and so on (see [POL 03], [POL 04], [POL 05]).

To get performance data over distance and generating some plots or tables like in [POL 05] for Telos platform in an outdoor environment, we tried to adapt a little of what was done in [ZHA 03] with Mica motes (indoor part of experiments).

Remember also that Telos uses the same Chipcon CC2420 radio transceiver (2.4 GHz wideband) as our Mica-Z, whilst older Mica employs RFM TR1000 (433 MHz narrowband), as in Table 7.2.

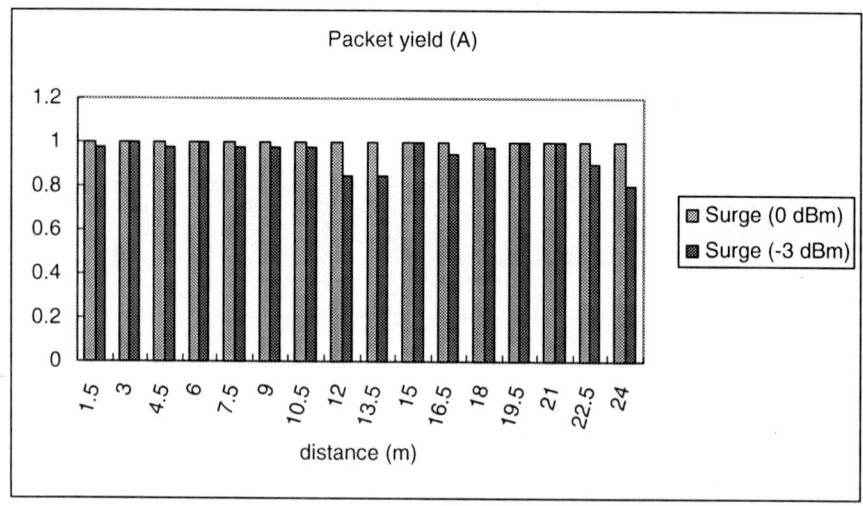

Fig. 7.11. Packet reception over distance with Surge

Using the experiment set-up described earlier, we found in general a good packet reception along the hallway, up to 34 m at 0 dBm (maximum output power for Mica-Z) when having the opportunity to go farther to the next building department.

Then power was decreased at −3 dBm to suit our environment better that is working over distances (∼25 m max. from base station) comparable to the expected range after the first test.

The results in both cases are showed in Figure 7.11, where left bars represent Surge values at 0 dBm.

Notice that at 34 m and 0 dBm output power, we once closed the entrance doors at the edges of the hallway, so that placing two obstacles between base station and remote mote, causing packet success rate (Yield field) to fall to ∼0.75. However, when operating Surge at −3 dBm, we encountered an odd behaviour between 11 m and 13 m, like "gray areas" cited in [ZHA 03]. In fact we experienced a very bad packet reception, falling to zero in certain cases, in that specific zone for samples taken in different test sessions.

In the following plot (Figure 7.12) there is a closer look at packet reception, zooming on measures between 9 m and 15 m at a finer granularity than we did before (x-axis labels not in scale).

We made the following hypotheses to explain these last results:

- Wireless communication uncertainty: we know that wireless channels are not deterministic, so it is normal to experience high variations in packet reception occasionally.

Practising

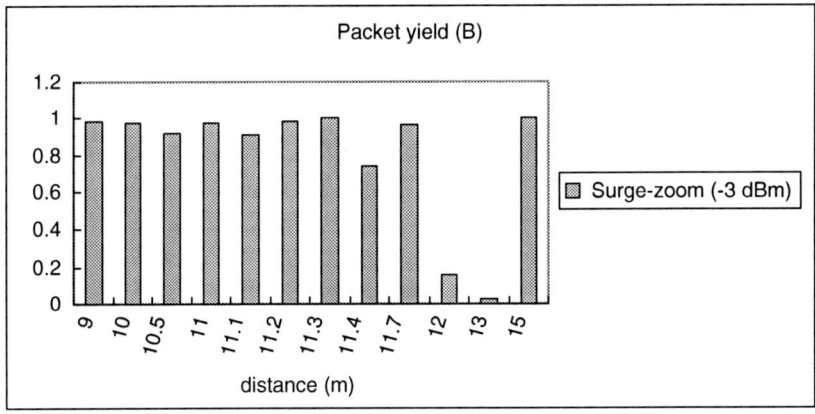

Fig. 7.12. Packet reception with Surge (zoom on "gray area")

- Interference from other devices in the 2.4 GHz band: even though we paid attention in setting up our testbed, following Crossbow's directions to avoid RF interference from Wi-Fi networks, we know that at least in one case there were colleagues working with IEEE 802.11b access points nearby.

- Surge behaviour: as we already said, this application is not the most straighforward one for transmitting/receiving packets only, because it runs multihop routing algorithms between motes and its complexity can lead to possible instabilities or freezing (at least from a user perspective)

To try to understand better what was going on in our environment, we then decided to use other applications/tools in order to confirm (or put even more doubts) the latest results.

The first thing that came up to our minds was to operate using "CntToLedsAndRfm" and "TOSBase" to generate traffic from a remote mote to a base station, and then back to a host PC. There is not much to say, in the sense that we did not find any "out-of-range" indication and reception was always good at the time of experiments, with packets sent from the remote mote well received at base station along the entire hallway. We could anticipate this result, because beyond the "gray area" packet reception with Surge was fine in many test sessions. However this further experiment confirmed at least this fact.

After that we tried to correlate values obtained so far with RSSI measurements taken along the corridor, adapting the Pong application for Telos to work with our Mica-Z software environment. A discussion on work done, together with plots of RSSI values over distance, is given next.

Practical Results – RSSI Measurements. As introduced early on, this test was designed to have an indication of signal coverage on our testbed, in order to pinpoint problems and possibly match what we referred as "gray area" of communication.

Some useful information on RSSI and LQI values are to be found in the datasheet of Chipcon CC2420, the IEEE 802.15.4 radio which comes with Mica-Z motes. Generally speaking, LQI is a measure of the "error" in the signal (at the demodulator), not the strength of the signal. Consequently a "weak" signal may still be very defined with no errors, and then deliver a good LQI over distance, as we can see if there is no interference for example from other 2.4 GHz devices.

Back to our issue, we found good values for RSSI and LQI, also between 11 m and 13 m, in two distinct test sessions (A and B) performed in different days and with two different pair of motes, to confirm and double-check measured data.

The Pong application collects signal strength and LQI value for mote at base station (always with ID = 1 for this application) and a remote mote, so it actually gives two symmetric measurements.

We can say that what we obtained is comparable with data seen in [POL 05] for Telos, in [XBO 05b] (Crossbow's MPR Manual) and typical values in Chipcon CC2420 datasheet. Plotted values are averaged over several hundred samples at each point of measurement (marks on the floor) with an RF output power of -3 dBm on ZigBee channel 26.

Extra-work could involve a check at 0 dBm (or even at a lower output power, e.g. -10 dBm), and a comparison with a "Path loss model", as in [ZHA 03] for Mica in the 900 MHz band. Moreover, we could not swap the position of base station with remote mote to look for asymmetries in our testbed, but as we saw Pong gives RSSI/LQI values in pair at least.

LQI is always high, around 106 and 107, as reported in Table 7.6 for session B.

As expected, RSSI decreases with distance, and received values at base station and remote mote follow the same trend in their fluctuations. The results for RSSI are showed in Figure 7.13.

The overall picture of RSSI values looks very similar in its trend. But more important we found that at the time of experiments we got always very good LQI data over distance that in the end is what interests more.

Besides what we have already said, the following other comments can be made:

- Differences in values between base station and remote station, even if they refer to symmetric points, can be explained with wireless communication channel characterization, i.e. space-time variant;

Practising 251

Table 7.6. Averate RSSI and LQI over distance – test session B

Distance (m)	At Base station		At Remote mote	
	RSSI (dBm)	LQI	RSSI (dBm)	LQI
1.5	−53.38	106.63	−55.63	107.10
3	−56.30	107.12	−56.19	107.30
4.5	−61.18	106.67	−63.97	107.05
6	−57.52	106.91	−61.19	107.26
7.5	−66.49	107.54	−69.08	107.69
9	−65.71	107.33	−69.04	107.44
10.5	−60.55	107.26	−63.51	107.44
12	−60.29	106.54	−63.54	106.88
13.5	−56.44	107.15	−59.89	107.08
15	−66.97	107.25	−69.79	107.06
16.5	−76.71	106.98	−79.18	107.18
18	−73.64	106.71	−76.91	106.91
19.5	−66.83	106.57	−69.99	106.67
21	−76.04	106.38	−78.81	106.67
22.5	−64.28	107.08	−67.00	107.26

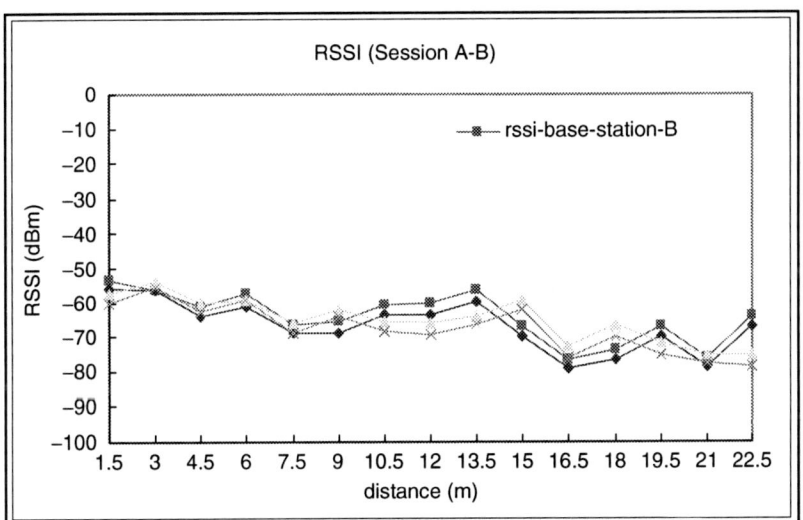

Fig. 7.13. Average Signal strength – test sessions A and B

- We can say the same thing about the fact that average values at one end (base station) are almost always higher than the others (remote mote) along the plotted graphs;
- We did not find a match with "gray area" we mentioned between 11 m and 13 m, so we suppose it was due both to interference from other devices in the 2.4 GHz band together with variations in the propagation channel;

Packet Reception Between Indoor Offices. This experiment is designed to check mote's communication in a real indoor environment composed of several offices. We will show some preliminary results for packet success and packet loss rate in a testbed composed of only two motes, moving from zone to zone of a building floor. We used the same "Surge" application we examined in previous experiments, but in a much simpler setup as we will detail better next.

Test Preparation and Layout. For this experiment we moved around the same floor where we did all the other tests. Then we chose our office lab as base station for collecting packets from different spots (zones or offices) and check for reception.

Figure 7.14 shows our basic set-up and different areas involved.

For this test we tried to simulate a typical deployment, without caring too much of mote's placement, provided that they were in the area to be checked. As noticed in the previous section, during the tests there were sometimes in progress some experiments with IEEE 802.11b equipment in an office nearby. However, here physical obstacles were the main limitation, as we will see next.

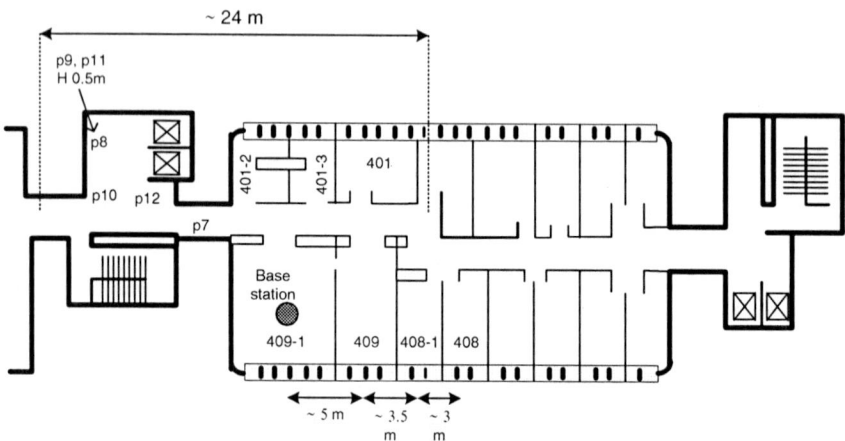

Fig. 7.14. Plan of the building floor with office labels

Methodology and Activities. There are many similarities with the two other experiments described so far. Here we are using just two motes in another context: starting from the first spot, the office lab marked 409-1, we moved the remote mote step by step, waiting to collect a certain number of samples before proceeding further to another point.

The experiment was repeated at maximum power and one level lower to get a couple of set of measures. Several log files were collected for statistical analyses and post-processing activities.

Experiment Set-up. The following parameters were chosen for the experiment, similar to what we did for multihop with Surge previously, but adapted for this specific context and then with higher power levels:

- Mote IDs: 0 and 77 for the pair of motes
- Mote Rf Output Power: 0 dBm and −3 dBm
- Surge Timer Rate: 2 (one packet every 2 s, maximum rate suggested for Surge)

Practical Results – Packet Reception. As said before, some results in an indoor environment can be found in [ZHA 03] in a different and more complex set-up involving several tens of Mica motes in the 433 MHz band. There the best indoor behaviour was with the lowest output power, because of shorter communication range between motes and less multipath interference.

Conversely, in our simpler testbed with just a couple of motes trying to transmit each other crossing walls and doors, we expect that the higher the power, the best the communication performance. In fact, repeating the experiment at different power level, 0 dBm and −3 dBm, we noticed a much better coverage at 0 dBm.

We can also say that we did not go very far in our indoor tests, just being able to reach one/two offices away from base station, depending on types of materials along the communication path.

For example, moving horizontally on the building floor depicted in Figure 7.14, walls separating adjacent offices have metallic surfaces which usually act as an RF shield. We suppose we could sometimes get over through the upper part of the wall (i.e. the ceiling).

Along the hallway instead we find generally structural panels in plastic or concrete and wooden doors. As a matter of fact, only crossing from one side of the building to the other was successful in most of the cases, but only at maximum power (0 dBm). The latter result is shown in Figure 7.15 below.

When operating Surge at −3 dBm (not showed in figure), we were not able to get any communication outside the office where we had the base station (lab room 409-1). However at that power we logged a file with multiple packet

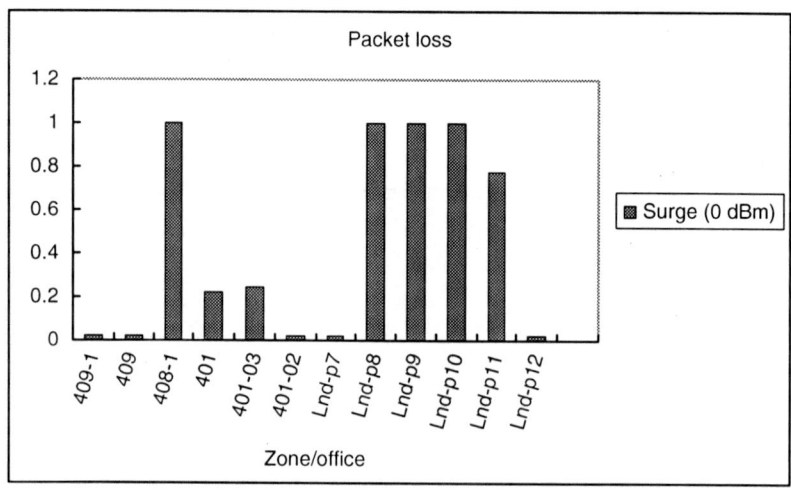

Fig. 7.15. Packet loss in indoor environment with Surge

reception, which we mentioned several times as due to multipath and signal reflections.

An example of that kind of Surge_reliable raw packets, as received on the PC control station, is showed below (output format adapted for readability, sample sets of multiple packets highlighted):

Node	Count	Seq	Hopcnt
77	15	210	1
77	16	211	1
77	17	212	1
77	18	213	1
77	19	213	1
77	20	213	1
77	21	213	1
77	22	214	1
77	23	214	1
77	24	214	1
77	25	214	1
77	26	214	1
77	27	215	1
77	28	215	1
77	29	215	1
77	30	215	1
77	31	215	1
77	32	215	1
77	33	216	1
77	34	217	1
77	35	218	1
77	36	219	1

Conclusions. At the end of the indoor test sessions we carried out (in an enclosed space, along a hallway, between offices), we can conclude that:

- Among all the experiments, the higher variability was found in the last test between offices, mainly because of physical obstacles along the communication path
- We experienced the same behaviour inside a single office as well, just moving a little bit one mote respect to the other
- In the latter case, it is mainly due to multipath, as seen in the log session shown previously
- When working in a small enclosed space, it is important to use the right amount of power, or to be general adjust it to the specific target environment
- Doing so permits to have a clearer communication channel for the intended topology, avoiding direct or indirect interference
- In a multihop testbed involving several contiguos offices and motes, probably we could make an application like Surge work (for what we have seen in our specific environment with also metallic panels, a high power level, i.e. 0 dBm, would be required to get over obstacles)

2. Practising WiMAX Equipment

In the previous chapters we have seen the concepts behind the new generation of WMANs and their technical specifications as a result of standardization bodies efforts led by the IEEE. In this section we are going to present some lab experiments employing one of the few certified WiMAX platform on the market.

The WiMAX product registry

The complete and updated list of all the manufacturers who went through the certification process endorsed by the WiMAX forum is available at the WiMAX Forum Certified™ Product Registry.[19] That provides specific information on Member's WiMAX Forum certified products, giving direct access to posted products and certifications by company or product category. As of the date, a list extracted from the online registry with company names and products is showed in the table right below just to provide a quick overview of the current trend.

---

[19] See: http://www.wimaxforum.org/kshowcase/view.

Since working with standards ensures specification compliance and interoperability of products from different vendors, it is certainly worth choosing from tested and certified products when viable. Sometimes proprietary solutions offer prompt response to particular customer demands, but in the long term, as with WLANs, operating within standards is a better and wiser choice.

Consider, for example, what recently is happening with telecom/cellular operators looking for the best alternative between 3G and other technologies in order to invest in the next future (Table 7.7). A wider availability of WMAN standard products, like WiMAX certified equipment, would clearly boost their

Table 7.7. Equipment records from the WiMAX Forum Certified Product Registry (Summer 2006)

Company	Product	Company	Product
Airspan Networks	ProST	Redline Communications	RedMAX Subscriber Station
Airspan Networks	MacroMAX	Redline Communications	RedMAX Base Station
Airspan Networks	EasyST	SEQUANS Communications	SQN2010-RD
Alvarion	Micro Base Station	SEQUANS Communications	SQN1010-RD
Alvarion	BMAX PRO-S CPE	SEQUANS Communications	SQN 1010-RD (FDD)
Alvarion	BreezeMAX Si	SR Telecom	Symmetry Base Station
Alvarion	Macro Modular Base Station	SR Telecom	SSU5000 Symmetry Subscriber Station
Aperto Networks	PacketMAX	Selex Communications	YSEMAX
Aperto Networks	PacketMAX 5000	Siemens SPA	WayMAX@vantage
Axxcelera Broadband Wireless	ExcelMAX FD CPE	Siemens SPA	Gigaset SE461 WiMAX
Axxcelera Broadband Wireless	ExcelMAX BS	Wavesat Wireless Inc.	Wavesat miniMAX 3.5 GHz (FDD)
Proxim Wireless Corporation	Tsunami MP16 3500	Wavesat Wireless Inc.	Wavesat miniMAX 3.5 GHz (TDD)

interest and encourage the adoption of IEEE 802.16 networks even in urban areas. Mobility is probably the last barrier to overcome, so the competition would become more real by the time IEEE 802.16e products would hit the market.

Equipment identification

The equipment we had the opportunity to try is produced by Redline Communications (http://www.redlinecommunications.com). This is a Canadian company founded in 1999 with headquarters in Markham, Ontario. As we can see in the last table, it is listed in the WiMAX Forum Certified Product Registry. From their web site we understand that the core technologies of their products, related to the data link layer (MAC), physical layer (PHY) and radio communications (RF), are internally developed. The product portfolio includes a wide range of radio equipment, operating in licensed and non-licensed spectrum mainly for fixed broadband wireless systems. Figures report over 20,000 installations in more than 80 countries, with a commercial network built around some of a hundred distributors.

The base station and subscriber station models are the following:

(a) RedMax AN-100U Base Station

(b) RedMax SU-O Subscriber Unit Outdoors

From the information on the certificate for the base station we obtain:

– Product name, model and type: RedMax, AN-100U, base station

– Issue date and number: January 2006, certificate #2

– Profile: 3.5 GHz (center frequency), 3.5 MHz (channel width), TDD (duplexing technique)

– Software version: 1.0.28

– Test laboratory: Cetecom, Spain

We can also notice that as of date we are at the very early stages of the overall process of bringing to market several interoperable products. As a curiosity, certificate #1 was issued to Aperto Networks base station the same day. Centro de Tecnología de las Comunicaciones (CETECOM) is a WiMAX Forum Designated Certification Laboratory (WFDCL) located in Spain, offering testing and certification services in wireless technologies.

As for profiles, a certification (Figure 7.16) for one profile does not imply that other profiles are not implemented on the equipment. For example a 7 MHz channel width option is available on the tested equipment as well.

Fig. 7.16. Redline RedMax product certificate

2.1 General Objectives and Action Plan

Back to our objectives, we will first describe the context and test scenario for the lab experiments we figured out. Then we will move on to practical aspects and details related to the WiMAX equipment in use, illustrating basic configurations and test results. We are interested in verifying sample configurations related to IEEE 802.16 specifications and profiles, and the extent to which current equipment implements them.

From a more technical perspective, a list of specific points we want to develop by our experiments is:

(a) To plan and implement a test environment for WMAN equipment

(b) To deploy and verify the operation of a simple point-to-point (PTP) or point-to-multipoint (PMP) network

(c) To make some performance measurements and preliminary tests

(d) To configure and test QoS and traffic priority capabilities (e.g. for voice or video streams)

2.2 Case Study: Redline WiMAX Equipment

The overall scope of this lab trial is to show a sample setup with WiMAX equipment. Our experiments are based on the RedMAX platform by Redline, operating in the 3.5 GHz licensed band. The product documentation coming with the equipment (see [RED 06a], [RED 06b], [RED 06c]) was a valuable source of information and very useful in helping understand better the IEEE 802.16 standard. The software version active on the base station was 1.0.53.

Practical Steps and Lab Setup. To reach our objectives some specific tasks are to be performed in a step-by-step fashion:

1 Equipment analysis and hardware overview

2 Study of product documentation and typical configurations

3 Preparation of plans for experiments and test checklists

4 Installation and configuration of all the necessary equipment and tools (hardware and software)

5 Collection and analysis of a set of repeatable measures

Figure 7.17 shows our basic network lab set-up, together with network addressing information.

A couple of things to be noticed are that this kind of equipment operates at open systems interconnection (OSI) layer 2, in other words they have no routing capabilities. Consequently, we would need some other external device, if layer 3 network segmentation were required.

Moreover, since we are using carrier-class equipment, some network services are thought to be provided by dedicated or centralized servers/devices, such as dynamic host configuration protocol (DHCP) and network time protocol (NTP) information for subscriber stations, as indicated in the network diagram.

The base station acts as a DHCP relay agent, forwarding DHCP requests to the configured DHCP server. The latter in turn can provide the DHCP client population all the expected information and options, including the address of an NTP server. A management connection identifier (CID) over the wireless link is used to carry these network exchanges.

Another important service to the base station or "sector controller", is that of time syncronization. On the AN-100U user's manual the preferred architecture is by an external global positioning system (GPS) clock reference. Many base stations (sector controllers) and a pair of GPS receivers can be connected together to provide redundancy of the time signal used for TDD operation. Using GPS time accuracy provides a more scalable and reliable architectures,

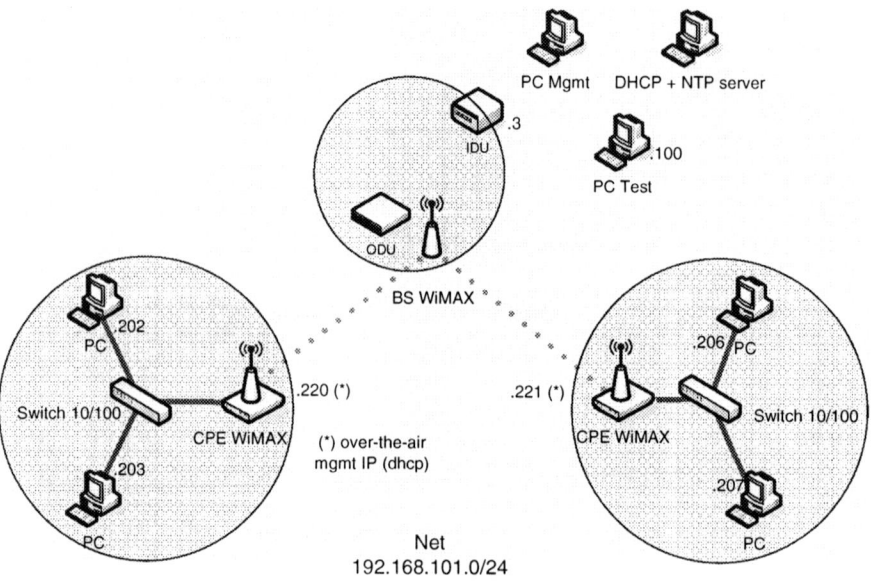

Fig. 7.17. Basic network diagram for WMAN equipment testing

when several base stations operate in close proximity using adjacent frequencies. Thus the use of the GPS clock minimizes inter-cell interference through accurate coordination of transmit and receive periods.

Base Station and Subscriber Unit Overview. The RedMAX AN-100U base station and RedMAX SU-O (subscriber unit or CPE) are IEEE 802.16-2004 compliant wireless devices for deployment of PMP and PTP systems.

Table 7.8 contains the available equipment we made use of at the time of experiments.

The base station as a whole consists of an indoor terminal (IDU) and outdoor transceiver and antenna (ODU). A complete wireless system is comprised of a base station and one or more subscriber stations. Each subscriber station (or CPE) registers and establishes a bidirectional data link with an authorized base station.

A base station functions as a central hub or concentrator, managing wireless links for remote subscriber stations and connecting this network to the outside world, typically by a WAN link.

In our tests, since we were operating in an indoor lab, it is important to limit the output power, especially at the base station. The maximum power from the system technical specifications is 23 dBm (region and regulatory domain specific), so we operated at 0 dBm with an additional external attenuator (20 dB loss).

Practising

Table 7.8. WiMAX equipment details

Equipment and accessories	Function	Qty	Notes
RedMax AN-100U base station	Indoor data unit – IDU	1	2 × Ethernet ports (data and management) 1 × IF port coaxial cable
Transceiver 24VDC, 1.5A	Outdoor data unit – ODU	1	1 × IF port coaxial cable 1 × RF port coaxial cable
Antenna 60° – 17 dBi	External sector antenna	1	Data taken from the Redline Link budget tool
RedMax SU-O subscriber unit outdoors	Outdoor unit	2	Integrated antenna (flat panel, 14 dBi)
POE – PowerDsine 3001	Power module (power-over-ethernet)	2	2 × RJ45 connectors

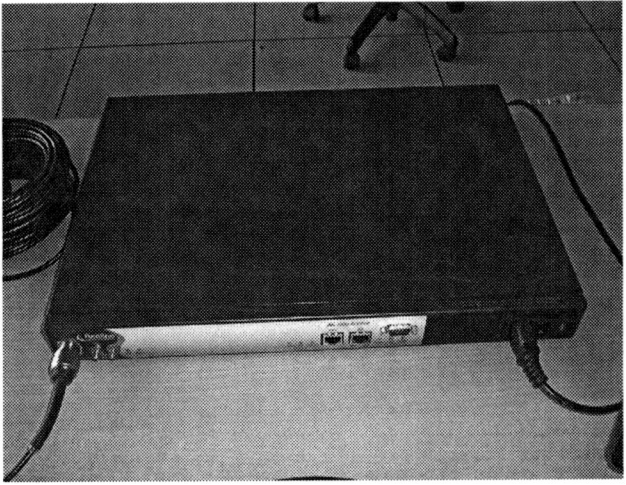

Fig. 7.18. Base station RedMAX AN-100U – BS IDU equipment

Some pictures of the equipment used in our test lab are shown in Figures 7.18 and 7.19.

As we can see, equipment at base station is comprised of several parts, depending on their functions and placement (indoor or outdoor). The AN-100U indoor unit (IDU) is the base station terminal. It performs all the management and control functions. Specifically, by its IF port, it sends/receives intermediate

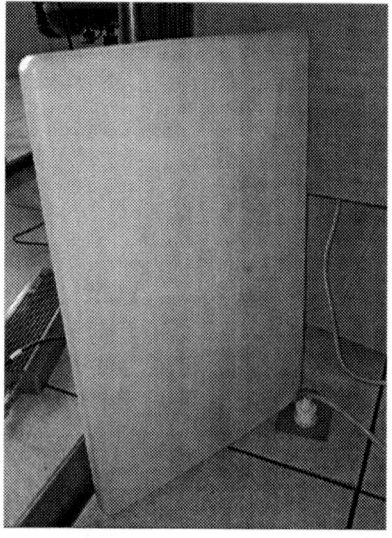

Fig. 7.19. Outdoor transceiver and antenna – BS ODU equipment

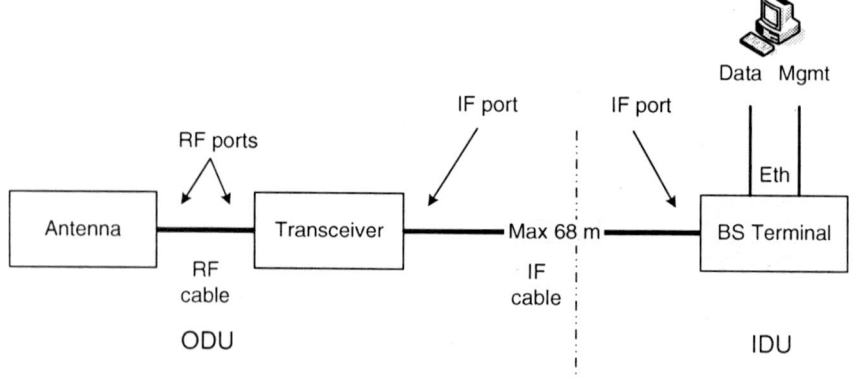

Fig. 7.20. Base station general connecting diagram

frequency (IF) modulated data to/from the transceiver, receives status information from the outdoor transceiver, transmits control information to the transceiver, supplies power to the transceiver.

The outdoor transceiver (ODU) is the radio transmitter/receiver operating at radio frequency (RF). It is the core radio system and is housed in a weatherproof aluminum alloy case. By its RF port, it sends/receives the actual RF signal to/from the antenna.

A general connecting diagram of all the base station equipment is shown in Figure 7.20.

Fig. 7.21. Subscriber station RedMAX SU-O and its power module (PoE)

Similarly the remote subscriber station SU-O (Figure 7.21) consists of indoor and outdoor parts: an outdoor radio (transceiver and antenna) and an indoor power-block with network connections. It is designed to ease the installation process, in a very straight forward way. It is sufficient to place each part according to its function and align the external unit to the intended base station. The unit includes an audible antenna alignment tool to assist in pointing the integrated antenna for maximum signal strength (the signal will sound infrequently when a low signal is detected, and more often as the signal strength increases).

It is also available an SU model (SU-ORF) with external mounting antenna, suitable for particular installations, such as when higher antenna gains are needed, as well as subscriber units for indoor placement only (SU-I).

Equipment Features and Technical Specifications. To complement the product analysis we have seen so far, it is important to decribe in short some key features and specifications of the tested equipment.

The AN-100U base station is compliant to WMAN–OFDM and WHUMAN–OFDM physical layer (PHY) profiles of the IEEE 802.16-2004 standard for 3.5 MHz and 7 MHz channelization (see discussion above on certification).

It is designed for 2–11 GHz operation of the physical layer definition, with supported frequency range between 3,400 and 3,600 GHz, using OFDM – 256 FFT for effective operation in harsh multipath environments.

This in an interesting point for outdoor deployments, since the intended system capabilities are for LOS, Optical LOS, Non-LOS fixed wireless access in a cell-based PMP scenario. Furthermore the RedMAX product does not support mesh communication (direct subscriber-to-subscriber).

As of date, the base station is reported hardware ready in order to provide encryption for user traffic based on the IEEE 802.16 MAC features and header fields, even tough the privacy sub-layer (e.g. for DES and AES) are *not available* in the current software revision.

One of the main objective of WiMAX is to provide a certain quality of service. The tested base station uses TDD to transmit and receive on the same RF channel. This is a non-contention based method for implementing an efficient and predictable two-way communication within a cell. The scheduling algorithm is managed by the WiMAX base station, that sends data traffic to subscriber stations, polls remote units for grant requests, and sends grant acknowledgements, based on the total traffic level, to all subscriber stations (SU).

To overcome the typical difficulties of the wireless channel, data transmitted over the wireless interface is protected using channel encoding techniques. The AN-100U base station supports convolution coding rates of 1/2, 2/3 and 3/4.

Other techniques are combined to maximize overall system performance: the ARQ mechanism, for automatic retransmission of errored and unacknowledged data without involving higher-layer protocols, and Reed Solomon error correction codes, for correcting bursts of errors.

The resulting data stream then modulates the OFDM signal carriers within the physical layer. The tested base station supports BPSK, QPSK and 16 quadrature amplitude modulation (QAM), and 64 QAM.

2.3   Lab Experiments

As anticipated at the beginning, we are interested in testing basic network performance and QoS capabilities (e.g. for voice or video streams) for a PMP fixed wireless access scenario.

Therefore using the lab setup illustrated in Figure 7.17 we want to configure and verify the operation of the WiMAX equipment described so far in the current section. We will also see snapshots of the web administration tool (Figure 7.22) available on the RedMAX base station, a very convenient way to control and understand better all the accessible product features.

Tools and Methodology.   To test basic network performance we are going to use some popular tools like "TrafGen,"[20] a Java based UDP and TCP traffic generator, and "Iperf,"[21] a client-server utility to measure TCP and UDP throughput performance.

---

[20] Refer to: http://www-lor.int-evry.fr/~vincent/java/trafGen/trafGenEn.htm.
[21] See: http://dast.nlanr.net/Projects/Iperf/.

Practising 265

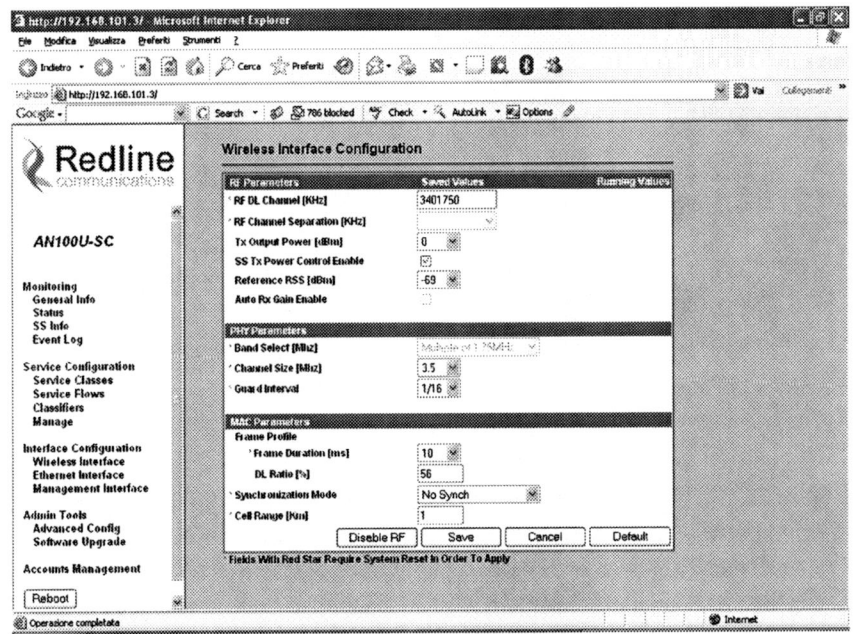

Fig. 7.22. Redline graphical web administration tool

We made use in particular of UDP traffic streams in order to inject a high packet rate and verify the effect on our sample network.

Typical measurements include maximum throughput, datagram loss, delay jitter depending, e.g. on packet size, data sending rate, traffic direction (upload/download).

Care has to be taken in choosing a neat set-up, at least to get started, such as to avoid fragmentation and have all the involved equipment and devices behaving predictably (e.g. far from being a bottleneck themselves). This obviously includes the machines used as traffic sources and sinks, which ought to be monitored for health/performance status during test sessions.

As for QoS measurements, the following general procedure was applied:

1 Set up of traffic sources with different priority requirements, such as "real time" (RT) high priority and "best effort" (BE) low priority

2 Injection of RT traffic flows in the network, like that originated by a multimedia source (see below), in order not to saturate link capacity

3 Addition of a traffic burst such to exceed wireless channel capacity (several Mbps), for example 20% higher, assuming that the bottleneck is there

4 Analysis of results at the sink, qualitative and quantitative, depending on the used tools

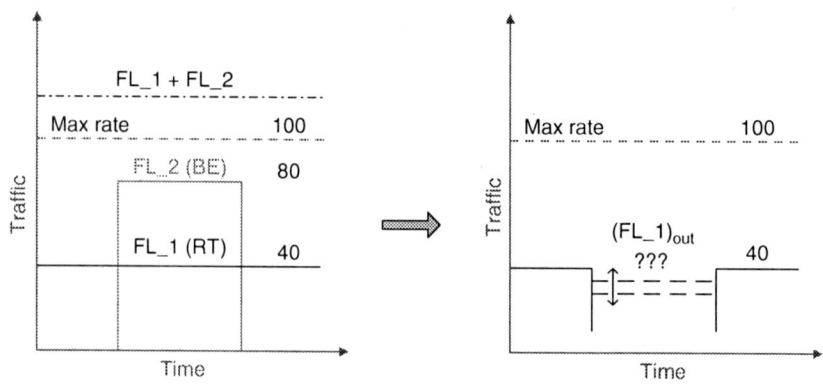

Fig. 7.23. Effect of two competing flows on the network (focus on FL_1, RT)

Figure 7.23 illustrates the concepts above for assessing QoS characteristics in a typical scenario with two flows having different requirements.

For multimedia traffic we made use of videoLAM client (VLC), a program freely downloadable from "http://www.videolan.org/vlc/". It can act, for example, as a motion picture experts group (MPEG) video source over IP located on the BS side, with one or more remote VLC player at CPE sides.

Configuration Details and Service Flows. One of the key aspects in configuring a WiMAX base station is related to service flow settings, a main feature of the IEEE 802.16 standard. A service flow represents a *unidirectional* data flow having separate QoS settings and priorities for uplink and downlink traffic, providing the ability to set up different connections to each served subscriber station.

The base station, controlling all uplink and downlink traffic scheduling, allows QoS settings enforcement for each service flow. By considering the requests of all subscriber stations, the base station schedules uplink and downlink traffic to conform to the required service level agreements.

To have a general idea of what is involved in creating service flows on the tested base station, these are the steps to follow with some notes aside:

(a) *Define one or more "Service Classes" (SC)*: this is a pre-requisite, since each SC defines a set of standard QoS parameters that can be associated with a service flow. By creating a set of standardized SCs, new service flows, that conform to them, can be added. Each SC definition includes traffic rates, latency settings, priority and transmission policy settings, as showed in Figure 7.24 below:

Practising

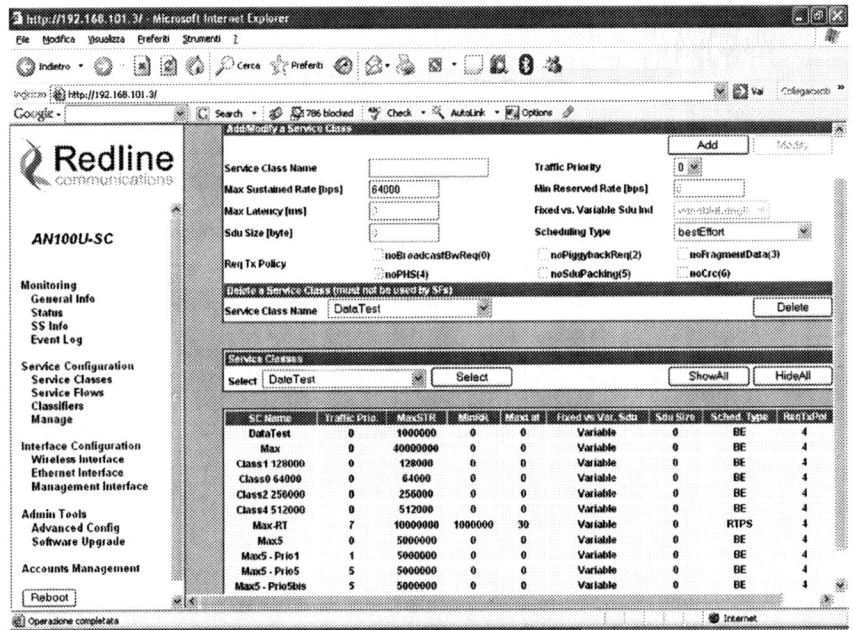

Fig. 7.24. Service Class configuration window on the base station web interface

Note: The current available scheduling types are BE and RT polling services (RT-PS). Other types, such as non-real-time polling services (nRT-PS)[22] and unsolicited grant services (UGS)[23] are not implemented as of the date.

(b) *Create one or more service flows (SF)*: all service flows are based on existing SC definitions, as seen above. Separate service flows are required for downlink and uplink traffic. Each definition includes identifying the subscriber station to which the SF applies, flow direction, class of service, and the classifier type (it refers to some network attribute, such as MAC addresses, as we will see next). The following figure shows the relative configuration window.

(c) *Define classifiers (CS) for each service flow*: a unique set of classifier rules can be defined for each service flow, depending on the classification type (i.e., packet or IEEE 802.3/Ethernet attributes) assigned when the service flow is created (see Figure 7.25). In Figure 7.26 we can see different classifier fields associated to an ID pair "SFID.ClsID", that is a combination of service flow ID and classifier ID.

---

[22] Suitable for example for FTP-like traffic.
[23] Suitable for example for VoIP-like traffic.

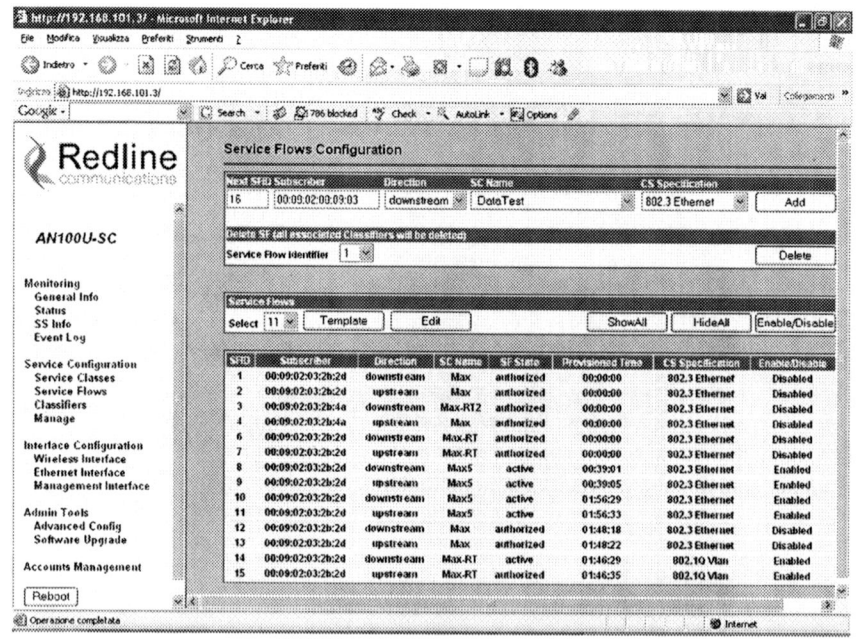

Fig. 7.25. Service Flow configuration window on the base station web interface

Defining the right classifier is the trickiest job in the overall configuration, because it is very easy to get confused along the process. Moreover it seems that some classifier is still not implemented or hard to make it work (e.g. trying to use VLAN tags to map a service flow).

The easiest set-up is by using MAC addresses to identify specific service flows, remembering always that there are other parameters involved related to SC and flow directions.

(d) *Activate all new classes of service, service flows and classifiers*: this is the final step to perform in order to bring to service the desired configuration.

As for VLAN mapping, it is important to review specific base station and subscriber unit features as well, since particular settings are needed in configuration of VLAN-aware devices like switches for tags, which travel across the network.

In the end, the base station centralizes all the configurations, controlling the operation of the wireless network and subscriber stations for almost every aspect (some particular CPE or SU feature for VLAN tagging, for example, are to be made directly via local command line inteface).

Practising

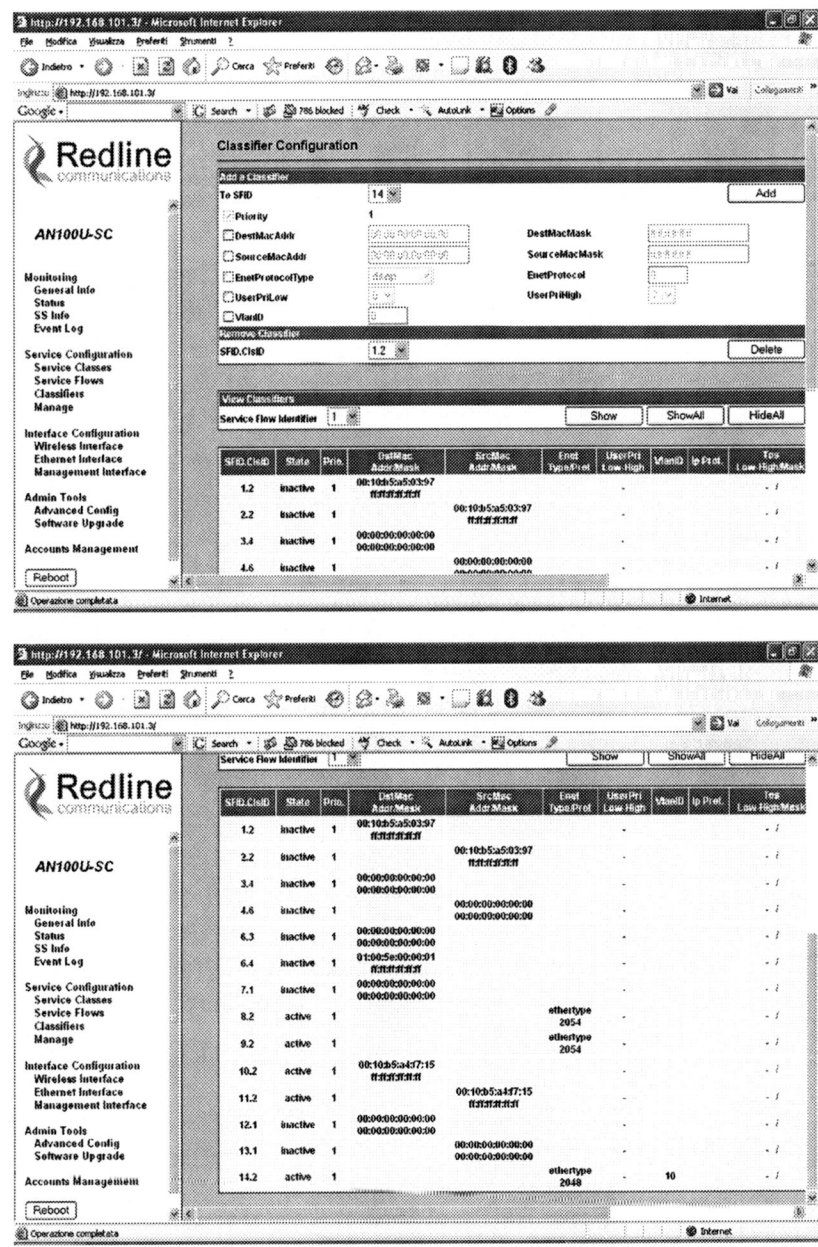

Fig. 7.26. Classifier configuration window on the base station web interface

Tests and Results. In order to understand and compare network performance in our lab set-up, it is useful to highlight some reference data coming from the technical specifications of the equipment under test (Table 7.9).

Table 7.9. Extract from Redline base station technical specifications

Channel bandwidth	Over the air bit rate (wireless PHY)	Ethernet bit rate (wired MAC)	Notes
3.5 MHz	Up to 17.5 Mbps uncoded rate[24]	Up to 10.8 Mbps max. Ethernet rate	*We assume this is for 64 QAM, minimum OFDM guard time (1/32) and code rate (3/4) on the wireless side*
7 MHz	Up to 35 Mbps uncoded rate	Up to 23 Mbps max. Ethernet rate	

Besides these best possible figures, on the same technical sheet it is correctly noted that actual Ethernet data throughput (not rate) is dependent on channel size, user protocols, packet size, burst rate, transmission latency and link distance. In other words, it depends on actual network usage type and traffic, as well as real context deployment.

As for latency, we see values ranging from 6 to 18 ms, depending on channel size and frame duration (see Figure 7.22 for examples of wireless interface configuration). This should be the minimum latency looking just at the wireless interface and its intrinsic transmission mechanisms.

UDP performance tests

Using the tools indicated before, we tried to reach and verify the maximum rates from the specifications. To generate a continuous UDP stream we made use of "TrafGen" with the following parameters and options:

1. Continuous emission with a relatively large datagram size (1024 bytes)

2. Downlink test (DL), running UDP send at BS side and UDP receive at remote side (SU or CPE)

3. Uplink test (UL), running UDP send at SU/CPE side and UDP receive at BS side

4. DL and UL test, running two parallel streams in opposite directions

The average received UDP results are in Table 7.10.

---

[24] This may be a typo on the specs sheet, since it is probably the coded rate (over the air bits).

Practising 271

Table 7.10. UDP results with TrafGen

Channel bandwidth	Downlink (DL)	Uplink (DL)	DL/max_rate %	Notes
3.5 MHz	5.8 Mbps	4.8 Mbps	53.7	DL ratio 56%
7 MHz	10.6 Mbps	9.6 Mbps	46.1	" "

The parameter "DL ratio" indicates the downlink usage as a percentage of frame size, providing traffic shaping on the uplink and downlink traffic profiles, and could become very important in setting the operating point of the network. It is accessible by the web administration interface of the base station (refer to "wireless interface configuration" window).

Based on the fact that the duplexing method is TDD, we can try to relate the difference between maximum expected data (from specifications) and values measured with TrafGen: at least in the first case with a 3.5 MHz channel bandwidth, the result for UDP payload is close to the maximum rate (around 10.8) multiplied by the DL ratio (0.56), meaning that the time division method affects the average throughput as one can expect. Even though we are not taking into account overheads and detailed differences between values in the two tables above, the result is pretty close since we are using large datagrams.

Similar measurements made with "Iperf" led to approximately the same results for maximum data streams.

In this case we can also get some other information related to "delay jitter" and "datagram loss" in the same test session easily. To verify the effect of different SC definitions, the same UDP tests were run with BE and RT flow schedules.

The average results with Iperf in different situations are shown in Table 7.11 with 3.5 MHz channel bandwidth, datagram size of 1,470 bytes, downlink direction.

The meaning of the SC parameters in the table are:

- MaxSTR: maximum sustained rate (also PIR) – a service flow created using this SC will be limited to sustained transmission at this rate (peak may be higher)

- MinRR: miminum reserved rate (also CIR) – a service flow created using this SC is guaranteed sufficient bandwidth for this rate

Those two elements were set at least equal to the input traffic rate for MaxSTR and higher than 1 Mbps for MinRR.

We can see that in this context the overall response is slightly better when using RT service flows with some sort of setting. The maximum UDP throughput

Table 7.11. UDP results with Iperf

	Service Class definition on BS					
	Best Effort @MaxSTR			Real Time @MinRR		
Input traffic[25] (Mbps)	Received throughput (Mbps)	Jitter (ms)	Lost datagrams (%)	Received throughput (Mbps)	Jitter (ms)	Lost datagrams (%)
1	1.00	7.2	0	1.00	7.2	0
2	2.00	5.7	0	2.00	4.5	0
3	3.00	4.1	0	3.00	2.9	0
4	3.99	3.2	0	4.00	2.3	0
5	5.00	3.1	0.024	4.99	1.9	0.12
6	5.84	3.2	1.4	5.87	3.5	0.87
7	5.87	3.3	15	5.86	3.6	15
8	–	–	–	5.87	1.8	26

is about the same as tested with TrafGen, that is around 5.8 Mbps in downlink (see comments before for comparisons with product specifications).

In general the delay jitter improves (i.e. decreases) with increasing input traffic; it is so up to around 5 Mbps, where maximum measured UDP throughput is approached and packet loss starts. The last value of jitter (i.e. 1.8) in the RT column is not significant, since losses are nearly 30%.

This result for jitter can be thought as a consequence of time division operations as before. Since transmission on the wireless channel is divided in fixed time frames, the best utilization would be filling up the available slot.

Intrinsic latency tests

In order to get an idea of transit delays, a very straightforward approach is to measure RTTs by ICMP packets with variable sizes, that is using "ping" and its extended options. Knowing at which point fragmentation occurs on the network allows us to choose reasonable lengths for test packets.

The setup is the same as before, with a PC at base station transmitting to a remote PC behind a CPE. This is for the moment the only traffic load on the network, so to not introduce perturbation.

The latency (RTT) results with different packet sizes are in the following table.

Latency is made up of several contributions, such as *queuing delay, transmit time* and *propagation delay*. Since test packets are sent out individually, waiting for a reply within a certain time before emitting another one, we can think on

---

[25] UDP bandwidth to send (-b option) in Iperf terminology.

Practising

Table 7.12. RTT results with extended Ping

Pkt length (bytes)	RTT min (ms)	RTT max (ms)	RTT mean (ms)	Length/max_rate (ms)	Lost Pkts (%)
32	42	219	92	0,024	0
64	37	253	93	0,047	0
128	38	264	96	0,095	0
256	38	234	89	0,190	0
1024	41	230	93	0,759	0
2048	39	229	82	1,517	0
4096	50	237	102	3,034	1
8192	61	211	100	6,068	0
16000	71	254	118	11,852	12
32000	103	259	145	23,704	35

the effects on each ICMP probe message. Looking at Figure 7.17, we can see that packets traverse the network starting from the PC connected to the BS, then go through the BS, the wireless medium, the CPE, the remote PC and then all the way back.

On the considered path to CPE, propagation delay is constant, whereas the other two factors may vary. In fact, transmit time changes with packet size, while queuing delay depends on various processing tasks and traffic load at nodes (devices).

Therefore, being constant all the other factors, longer probe packets have higher delays, as we can see in general in Table 7.12. When data packets are greater than 1,500 bytes fragmentation occurs, imposing assembly/disassembly of Ethernet frames carrying ICMP probes. The base station receiving these fragments may perform a packing operation (see chapter 5), gathering multiple MAC SDUs into a one MAC PDU. The maximum length of an IEEE 802.16 MAC PDU is 2,048 bytes, so going up with packet size will probably occupy more than one transmit frame, increasing system latency as well.

Just as an exercise, in the same table above we have indicated an approximate transmit time contribution at BS, calculated as if we had a link working at "max_rate" speed (Max Ethernet rate from the BS specifications) without considering fragmentation.

Note: Trying to ping the remote CPE radio interface, instead of the lab PC behind it, by its over the air management IP (see Figure 7.17) exhibits very high round trip delay. We can explain this fact considering that in general ICMP packets have very low priority in the processing operations of network equipment (this is especially valid for routers), therefore precious board resources (e.g. memory, CPU, interfaces) are used to perform relevant tasks before any other one.

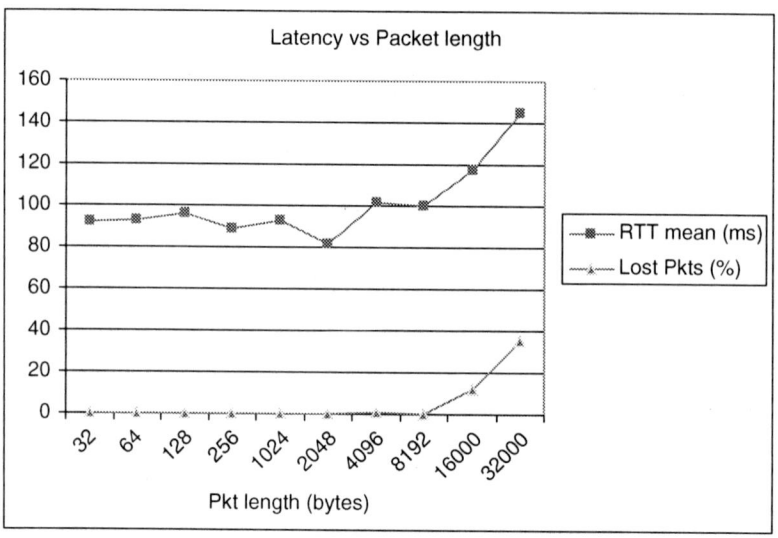

Fig. 7.27. Network latency and packet loss results

From Figure 7.27 we can notice that up to a certain packet length (around 8 KB), network latency, here RTT, lies between 80 and 100 ms. We can suppose this is an area dominated by internal processing of the involved equipment, mainly independently on packet sizes.

Then further increases in data size tend to increase losses and delay times as well in a more direct way. This may be due to the increasing contribution of transmit time, no more negligible in respect to other factors and direcly related to the amount of data to be transmitted.

As for lost packets, in general the longer the packet, the more the probability for communication errors to occur. Since we are just probing the network with ICMP packets, there is no way to recover communications error over a certain threshold by means of higher protocol layer mechanisms. In this case, considering also fragmentation in the network, if just a fragment is lost on some link in the path, the entire original packet is lost.

As seen in chapter 5, the IEEE 802.16 standard has an ARQ mechanism to provide automatic retransmission of errored data and unacknowledged data on the wireless channel without involving higher layer protocols. This helps a lot in containing protocol overheads and delays, but cannot stand any impairment, as well as things happening in other parts of the network.

QoS bandwidth tests

Using the same methodology indicated at the beginning of this section and ideas illustrated in Figure 7.23 for QoS measurements, our aim is to verify

Practising

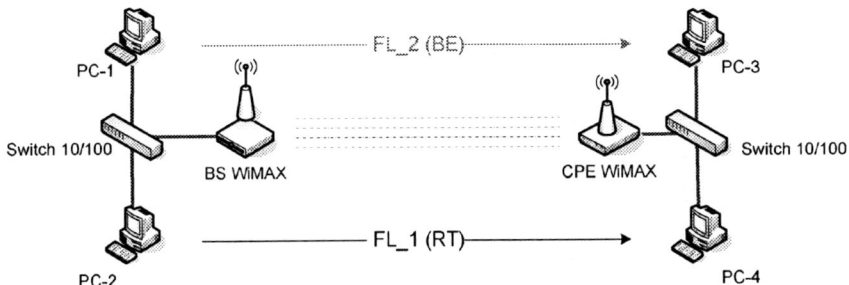

Fig. 7.28. Lab setup for QoS tests

whether we can maintain an RT traffic (e.g. multimedia) on the network when other competing flows or peaks at lower priority (BE) are trying to use the same resources.

The tricky part here is to configure service flows properly, a task which requires careful attention as we have seen previously. Not every detail was clear from the product documentation available to us, there might be something more specific for configuring this features, such as some application notes. Anyway in the end we found a working configuration using MAC addresses as classifiers and splitting flows requiring different QoS onto two parallel test equipment, as shown in Figure 7.28.

It is useful to remember that a connection in WiMAX is a unidirectional mapping between base station and subscriber station MAC peers for transporting service flow's traffic. As seen in chapter 5, a 16-bit CID identifies connections. All traffic is carried on a connection, even for service flows that implement connectionless protocols, such as IP. A CID maps to a service flow identifier (SFID), which defines the QoS settings of the service flow associated with that connection.

Following the procedure highlighted at the beginning of this section to configure service flows, hereafter are reported the performed steps and some details:

1. Definition of SC for BE and RT type traffic

    (a) *BE class*: SC name = Max, MaxSTR = 40 Mbps, traffic priority = 1, scheduling = best effort;

    (b) *RTPS class*: SC name = Max-RT, MinRR = 1.5 Mbps, traffic priority = 7, scheduling = RealTimePollingServices;

2. Configuration of service flows for uplink and downlink traffic: two service flows (UL + DL) per traffic type (BE and RTPS) are required for the same Subscriber Station (SS), resulting in four active SFs:

(a) Service flow 1: SFID = 1, subscriber = <CPE MAC address>, direction = downstream, SC name = Max, CS specification = IEEE 802.3 Ethernet

(b) Service flow 2: SFID = 2, subscriber = <CPE MAC address>, direction = upstream, SC name = Max, CS specification = 802.3 Ethernet

(c) Service flow 3: SFID = 3, subscriber = <CPE MAC address>, direction = downstream, SC name = Max-RT, CS specification = IEEE 802.3 Ethernet

(d) Service flow 4: SFID = 4, subscriber = <CPE MAC address>, direction = upstream, SC name = Max-RT, CS specification = IEEE 802.3 Ethernet

3. Definition of classifiers to map created service flows;

   (a) Classifier 1: SFID.ClsID = 1.1, DestMAC = <MAC PC-3>, Dest-MAC Mask = ff:ff:ff:ff:ff:ff

   (b) Classifier 2: SFID.ClsID = 2.1, SourceMAC = <MAC PC-3>, SourceMAC Mask = ff:ff:ff:ff:ff:ff

   (c) Classifier 3: SFID.ClsID = 3.1, DestMAC = <MAC PC-4>, Dest-MAC Mask = ff:ff:ff:ff:ff:ff

   (d) Classifier 4: SFID.ClsID = 4.1, SourceMAC = <MAC PC-4>, SourceMAC Mask = ff:ff:ff:ff:ff:ff

4. Generate traffic with real time requirements from PC-2 to PC-4 and with best effort requirements from PC-1 to PC-3;

Note: The "traffic priority" parameter indicated above for SCs is not related to priority settings defined in Ethernet 802.1p. It is the priority to be used for service flows created using that particular service class. This priority is relative only to other service flows on the same subscriber station, where a value of 7 represents the highest priority.

To get quantitative measurements, we made use of the same tools described earlier, in particular "Iperf", which allows generating tunable UDP traffic streams easily.

The results with different flow combinations are in Table 7.13 (refer to Figure 7.28).

From the results above and other tests carried out, we have noticed a general good response for expected bandwidth on the QoS stream by defining two completely different service flows, related to BE and RT service classes.

Outcomes were different trying to set different concurrent BE flows having a combined MaxSTR lower than the maximum seen in previous tests (e.g. about 5.8 Mbps on downlink).

Table 7.13. QoS results with different flows

Test case	FL_1 (RT)	FL_2 (BE)	Received stream (PC-3)	Received stream (PC-4)
A	On	Off	1.5	NA
B	Off	On	n.a.	5.8 approx.
C	On	On*	~1.5	Less than the remaining capacity (3.6) and pkt loss

*Perturbing burst, up to 6 Mbps at the source.

For example, considering the same lab set-up as before we found weird behaviours depending on priority settings and order of transmission:

- *Case 1*: two service flows (uplink + downlink) with BE class of service to serve any PC behind the CPE (by entering DestMAC Mask = 00:00:00:00:00:00);
  in this case PC-3 and PC-4 share the available bandwidth and none of them gets more than the other

- *Case 2*: two service flows per direction and per PC with BE class of service to match PC-3 and PC-4 MAC addresses;
  in this case we have a total of four service flows as before for BE and RT, but we noticed that one of them got always priority over the other; for example, generating two flows of 3 Mbps to PC-3 and PC-4, one of them receives 3 Mbps with 0% loss, the other gets just a couple of Mbps with 30% loss; increasing the traffic to the more favoured PC up to the maximum allowable on the link, causes the second one to stop receiving and have 100% loss;

We suppose this behaviour is either due to an erroneous configuration (even tough using just MAC addresses it should be quite straightforward) or to an incoherent functioning on the base station. It could also come out from service flows having the same "traffic priority" (see before), even if mapped to different MAC addresses (classifiers).

QoS qualitative tests – VoIP streams

To verify VoIP communication using the same set-up we have just seen for QoS bandwidth measurements, it is sufficient to configure FL_1 (RT) to support VoIP traffic (MinRR > 64 Kbps) and FL2_2 (BE) as before.

Then looking at Figure 7.28, let PC-1 and PC-3 be two SIP phones (soft or hard VoIP phones). Consequently to establish and test a voice communication, it is just a matter of calling the VoIP peer and compare its perceived quality (e.g. clearness, fluidity, etc.) while adding a second flow (BE) at increasing rates.

Following this simple procedure and the same tools as before, we have been able to establish a satisfying VoIP communication using the QoS features and service flows for RT and BE traffic. Without activating QoS mechanisms, increasing the parallel BE stream by means of "Iperf" leads rapidly to degraded performance in voice communication.

In this test session we just experienced sound quality directly, but we did not mention some important factors for real time traffic, such as jitter and network delay. The latter, for example, has to be less than 150 ms round trip for a good conversation quality.

However, when traffic flow generated for FL_2 (BE) exceeds 10 Mbps, despite of the fact we had some QoS settings on the network, we observed degradation on the first RT flow as well. It is probably so since we are approaching the maximum Ethernet rate on the base station that is about 10.8 Mbps from its specifications (see Table 7.9).

QoS qualitative tests – video streams

Following the same procedure we have just seen for VoIP traffic, it is possible to test the reception of a video stream using VLC, a software for multimedia traffic. For example we used it as an MPEG video source over IP located on the BS side, with one or more remote VLC player installed on PCs behind subscriber stations.

Then configuring the different service flows for RT and BE traffic as before, with a MinRR of 1.5 Mbps for RTPS class of service, we were able to send a multimedia stream across the wireless network meeting the required quality.

As with VoIP tests, the parallel perturbing flow has to be no more than 10 Mbps to not incur in severe degradation caused by congestion at base station. In that case, even with QoS mechanisms in place, correct network operation is disrupted.

Since it is possible to set VLC for multicast streaming, we tried to configure the base station and its classifiers to manage this type of traffic cleverly. As already seen, configuring classifiers can easily become a challenging task. Some internals are not clearly documented, so at first we made use of MAC addresses to map service flows. Then some checks were made in order to understand better how to deliver multicast traffic properly, such as working with IPv4 or VLAN (IEEE 802.1Q) type classifiers (these are defined when creating service flows, see above). Since it did not perform as expected, we went back to Ethernet 802.3 style classifiers to make it work at least for basic tests.

It is worth noticing that multicast traffic is transparent to base station and needs to be classified as other flows by classifiers to get it through the network. This is in contrast to other layer-2 devices, such as network switches, for which multicast frames are treated as broadcast in general. We did not find any mention about "IGMP snooping" support for the tested equipment, so probably multicast management is a function that would be requested to other equipment (like routing and other features discussed at the beginning of this section).

Conclusions. At the end of all the tasks and tests we carried out with recent WiMAX equipment, we can conclude that:

- The tested equipment manufactured by Redline actually implements the IEEE 802.16 standard discussed in chapter 5

- As for the software version coming with the base station we used (release 1.0.53) some features are not implemented yet, such as for security and scheduling types (nRT-PS, UGS)

- Nonetheless a lot of features and a useful administration tool are readily available for building PTP and PMT WiMAX networks

- Even if operating only in a lab environment for analysis and verification of equipment, our basic tests helped to understand better the IEEE 802.16 standard and compare data with real product specifications

- Performance tests were made for UDP streams and multimedia traffic, getting some interesting data on downlink/uplink throughput, jitter, latency and qualitative VoIP/Video communication

- Some limitations on the equipment concern configuration awareness for classifier's use and layer-2 only capabilities

- QoS features are available and performed tests showed good results running competing flows on the network with different requirements (BE and RT)

There is a lot of extra work to do to characterize completely even the simple network used for our basic tests and experiments. This might include a combination of flows and traffic with variable size packets and statistic, as well as topologies with two or more subscriber stations. Another limitation is in the use of available general purpose hardware and programs. To get deeper in testing specialized platforms and commercial tools are on the market, but considering the required investments this is typically a job for dedicated test labs.

Appendix A: Structure of IEEE 802.11 Packets at Various Physical Layers

This appendix gives a detailed description of the structure of packets in IEEE 802.11 for the different physical layers.

1. Packet Format of Frequency Hopping Spread-spectrum Physical Layer (FHSS PHY)

The packet is made up of the following elements (Figure A.1):

1.1 Preamble

It depends on the physical layer and includes:

- *Synch*: a sequence of 80 bits alterning 0 and 1, used by the physical circuits to select the correct antenna (if more than one are in use), and correct offsets of frequency and synchronization.
- *SFD*: *start frame delimiter* consists of a pattern of 16 bits: 0000 1100 1011 1101, used to define the beginning of the frame.

1.2 Physical Layer Convergence Protocol Header

The Physical Layer Convergence Protocol (PLCP) header is always transmitted at 1 Mbps and carries some logical information used by the physical layer to decode the frame:

- *Length of word of PLCP_PDU* (PLW): representing the number of bytes in the packet, useful to the physical layer to detect correctly the end of the packet.
- *Flag of signalization PLCP* (PSF): indicating the supported rate going from 1 to 4.5 Mbps with steps of 0.5 Mbps. Even though the standard gives the combinations of bits for PSF (see Table A.1) to support eight different rates, only the modulations for 1 and 2 Mbps have been defined.
- *Control error field* (HEC): CRC field for error detection of 16 bits (or 32 bits). The polynomial generator used is $G(x) = x^{16} + x^{12} + x^5 + 1$.

The number of operating channels must be greater than or equal to 20 and less than 35 slots of 1 MHz in the band indicated in Table A.2.

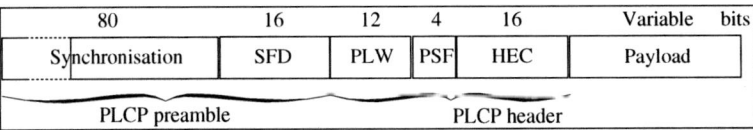

Fig. A.1. FHSS 802.11 packet

Table A.1. Supported rates

PSF bit 0	PSF bits 1-2-3	Rate (Mbps)
0	000	1.0
0	001	1.5
0	010	2.0
0	011	2.5
0	100	3.0
0	101	3.5
0	110	4.0
0	111	4.5

Table A.2. FHSS channels

Lower limit	Upper limit	Legal coverage
2.448 GHz	2.482 GHz	2.4465 – 2.4835 GHz

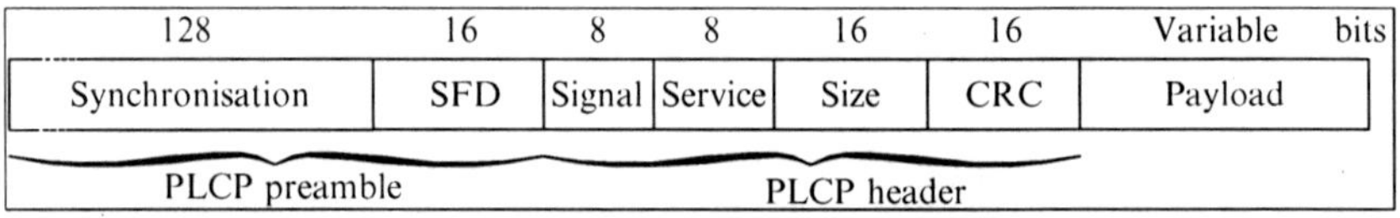

128	16	8	8	16	16	Variable	bits
Synchronisation	SFD	Signal	Service	Size	CRC	Payload	

PLCP preamble | PLCP header

Fig. A.2. DSSS 802.11 packet

2. Packet Format of Direct-Sequence Spread Spectrum Physical Layer (DSSS PHY)

The packet has the following structure (see Figure A.2).

2.1 Preamble

It is dependent on the physical layer and includes:

- *Synch*: it is composed of 128 bits set to «1» and cooperates to the synchronization of the receiver.
- *SFD*: *start frame delimiter* consists of a pattern of 16 bits: 1111 0011 1010 0000, used to define the beginning of the frame.

The preamble must be transmitted at 1 Mbps with a DBPSK modulation.

2.2 PLCP Header

The PLCP header is always transmitted at 1 Mbps. It contains several fields:

- *Signal*: this field allows to specify the modulation to use for reaching the desired rate, having 0A for 1 Mbps with a DBPSK modulation and 14 for 2 Mbps with a DQPSK modulation.
- *Service*: this field is reserved for future use and must be initialized to 0. A value of 00 signifies that the unit conforms to the IEEE 802.11 standard.

Appendix A

- *Length*: this field indicates the size of the frame in bytes, ranging from 4 to 8192 bytes.
- *Control error field* (CRC): CRC field for error detection of 16 bits (or 32 bits). The polynomial generator used is $G(x) = x^{16} + x^{12} + x^5 + 1$.

3. Packet Format of IEEE 802.11b HR/DSSS PHY

The packet structure of IEEE 802.11b DSSS is illustrated in Figure A.3.

Notice the presence of two preambles, one similar to that of the IEEE 802.11 standard with 128 bits set to 1 (scrambled bits) and another one shorter with 56 bits set to 0 (scrambled bits) that reduces the overhead.

Next we are going to describe just the fields which are different from those we have already seen in section 2:

- *SFD*: it allows the receiver to find the beginning of the frame. This 2-byte field is represented by the sequence 1111 0011 1010 0000 in the case of a long preamble, and its opposite 0000 1100 0101 1111 in the case of a short preamble.
- *Signal*: this field, in the case of a long preamble, allows to specify the modulation to use for reaching the desired rate, having 0x0A for 1 Mbps with a DBPSK modulation, 0x14 for 2 Mbps with a DQPSK modulation, 0x37 for 5.5 Mbps with a CCK4 modulation and 0x6E for 11 Mbps with a CCK8 modulation. With a short preamble, only three rates are possible (2, 5.5 or 11 Mbps).

IEEE 802.11b considers the utilization of a method to rearrange MAC frames after the CRC calculation phase.

The preamble must be transmitted at 1 Mbps with a DBPSK modulation. In order to reduce the overhead time contribution, the header in the case of a packet with a short preamble is transmitted at 2 Mbps using a DQPSK modulation.

Long packet

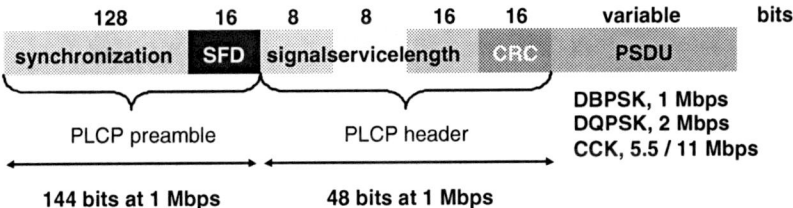

Short packet

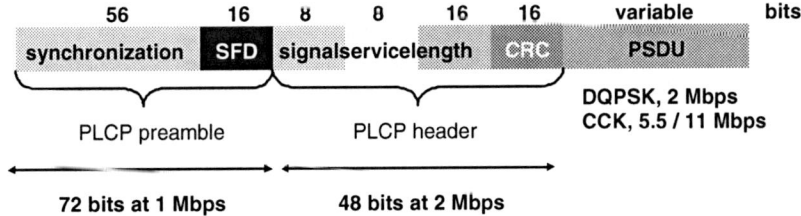

Fig. A.3. DSSS 802.11b packet

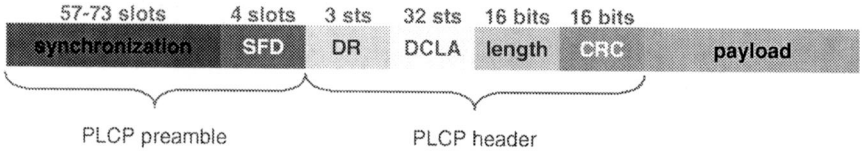

Fig. A.4. IEEE 802.11 IR packet

4. Packet Format of Infrared Physical Layer (IR PHY)

The name of the modulation in use is *pulse position modulation* (PPM). It is based on the IEC 60825-1, ANSI Z136 standard (Figure A.4). Different data rates are offered:

– Mbps with a 16-PPM modulation
– Mbps with a 4-PPM modulation

The packet is composed of the fields described afterwards.

4.1 Preamble

It contains:

– *Synch*: sequence of presence/absence of pulses in a series of slots.
– *SFD*: *start frame delimiter* consists of a pattern of 4 bits 1001 (1 pulse in a slot L-PPM and none for 0), used to define the beginning of a frame. A slot corresponds to one of the L positions of the symbol and has a duration of 250 ns.

4.2 Physical Layer Convergence Protocol header

It includes the following fields:

– *DR* (*data rate*): this field indicates the actual rate (000: 1 Mbps; 001: 2 Mbps). In the case of a transmission at 1 Mbps, the 16-PPM modulation is used (basic rate). The 4-PPM modulation is used in the case of a rate of 2 Mbps.
– *DCLA* (*DC-level adjustment*): this field is needed for letting the receiver establish the DC level after the reception of the fields SYNC, SFD and DR. It contains a sequence of 32 slots (1 Mbps: 00000000 10000000 00000000 10000000, 2 Mbps: 00100010 00100010 00100010 00100010).
– *Length*: this field indicates the number of bytes of the PSDU to transmit.
– *Control error field* (CRC): CRC field for error detection of 16 bits. The polynomial generator used is $G(x) = x^{16} + x^{12} + x^5 + 1$.

The main parameters of the different physical layers are summarized in Table A.3. They can be tuned to reduce time delays and adapt to environment conditions.

5. Packet Format of OFDM PHY (Physical Layer of IEEE 802.11a)

The packet structure considered at the physical layer level includes, like the other packets, a synchronization part and a header, indicating the modulation type in use and the transmission characteristics to the MAC layer. Figure A.5 clearly shows this structure.

Appendix A

Table A.3. Main parameters

Parameter	FHSS	DSSS	HR/DSSS	IR
Slot Time (microsec)	50	20	20	8
CCA Time (microsec)	27	<=15	<=15	5
Preamble (microsec)	96	144	144(**)	16/1 Mbps 20/2 Mbps
PLCP header (microsec)	32	48	48	41/1 Mbps 25/2 Mbps
MPDU max size (bytes)	4095(*)	8192	4095	2500
Modulation	GFSK	DBPSK DQPSK	DBPSK DQPSK CCK	PPM
Spectrum	Regulation restriction	Regulation restriction	Regulation restriction	No restriction

* Recommended value: 400 bytes – 1 Mbps, 800 bytes – 2 Mbps; this corresponds to a frame <3.5 ms.
** 144 is the preamble size for a packet with a long header, a packet with a short header has a preamble of 72 leading to an overhead of 192 µs in the first case and 96 µs in the second case.

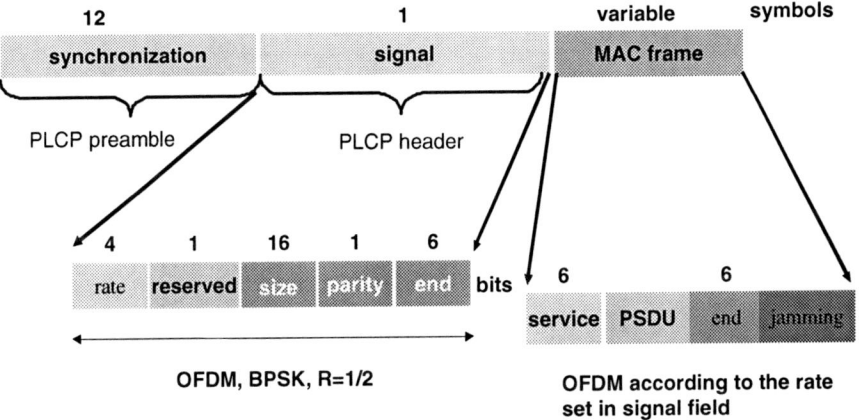

Fig. A.5. IEEE 802.11 OFDM packet

5.1 Preamble

It contains 12 OFDM symbols to perform the synchronization between the transmitter and the receiver. The duration of this preamble is of 20 µs. The synchronization is divided into two parts. The first part consists of sending ten OFDM training symbols lasting 0.8 µs. These symbols are sent over 12 carriers. The second part starts with a guard interval, lasting 1.6 microseconds, followed by two OFDM training symbols lasting 3.2 µs. These two symbols use the 52 available carriers and the BPSK modulation.

Table A.4. Available data rates

Rate field	Data rate (Mbps)
1101	6
1111	9
0101	12
0111	18
1001	24
1011	36
0001	48
0011	54

Table A.5. Data rates and modulations

Data rate	Modulation and coding rate	Coded bits per carrier*	Coded bits per OFDM symbol	Data bits per OFDM symbol**
6	BPSK, $R = 1/2$	1	48	24
9	BPSK, $R = 3/4$	1	48	36
12	QPSK, $R = 1/2$	2	96	48
18	QPSK, $R = 3/4$	2	96	72
24	16QAM, $R = 1/2$	4	192	96
36	16QIAM, $R = 3/4$	4	192	144
48	64QAM, $R = 2/3$	6	288	192
54	64QIAM, $R = 3/4$	6	288	216

* The number of coded bits depends on the modulation method (BPSK, QPSK, QAM).
** The number of bits per symbol depends on the rate of the convolution code.

5.2 PLCP Header

Lasting 4 µs, this header is transmitted across the "signal" field and contains the following fields (Table A.4):

- *Rate*: this field over 4 bits indicates the actual data rate.
- *Length*: indicates the size of the frame, protected by a convolution code.
- *Parity bit and reserved bit*: bit 4 is reserved for future use and is set to 0. The parity bit applies to the first 16 bits of the "signal" field to protect them against errors.
- *Tail*: end field over 6 bits (000000).

Contrarily to the other fields in the header, the service field is transmitted within the data part of the PDU as an OFDM symbol of 3.2 µs with a guard interval of 0.8 µs. The first 6 bits are set to 0 and assist to initialize the scrambler. The applied modulation is BPSK at 6 Mbps.

The coding scheme used for the data depends on the rate. The fields «tail» and «pad» are used in a way so that the total length is an integer multiple of the length of a block. In addition, the length of a block depends on the modulation and the coding scheme.

Table A.5 summarizes the modulation methods and the coding scheme employed for each of the supported data rates for the OFDM PHY layer.

Appendix B: IEEE 802.11 MAC Frames Structure

This appendix gives a detailed description of the structure of frames transmitted at the MAC layer level.

1. Different Types of MAC Frames

Table B.1 summarizes the type of frames found at the MAC layer.

Table B.1. MAC frames

Type Value (b3b2)	Type Description	Subtype Value (b7b6b5b4)	Subtype Description
00	Management	0000	Association request
00	Management	0001	Association response
00	Management	0010	Reassociation request
00	Management	0011	Reassociation response
00	Management	0100	Probe request
00	Management	0101	Probe response
00	Management	0110-0111	Reserved
00	Management	1000	Beacon
00	Management	1001	ATIM
00	Management	1010	Disassociation
00	Management	1011	Authentication
00	Management	1100	Deauthentication
00	Management	1101-1111	Reserved
01	Control	0000-1001	Reserved
01	Control	1010	Power save-poll (PS-Poll)
01	Control	1011	RTS
01	Control	1100	CTS
01	Control	1101	ACK
01	Control	1110	CF-End
01	Control	1111	CF-End + CF-ACK
10	Data	0000	Data
10	Data	0001	Data + CF-ACK
10	Data	0010	Data + CF-Poll
10	Data	0011	Data + CF-ACK + CF-Poll
10	Data	0100	Null function (no data)
10	Data	0101	CF-ACK (no data)
10	Data	0110	CF-Poll (no data)
10	Data	0111	CF-ACK + CF-Poll (no data)
10	Data	1000-1111	Reserved
11	Reserved	0000-1111	Reserved

2. Management Frames

The MAC header is identical for all the management frames (Figure B.1):

- *BSSID* is the address of the access point in the case of a network of type infrastructure or the identifier IBSS in an ad hoc network.

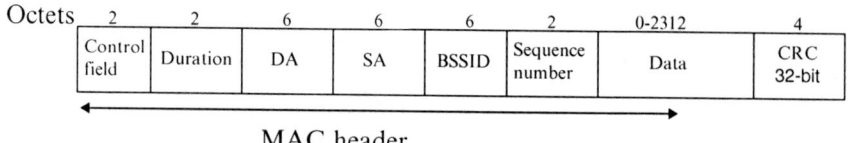

Fig. B.1. Format of a management frame

Table B.2. Structure of a beacon frame

Order	Information
1	Temporal timestamp
2	beacon Interval
3	Supported functions
4	SSID
5	Offered throughputs
6	FHSS parameters set
7	DSSS parameters set
8	CF (mode PCF) parameters set
9	IBSS parameters set
10	TIM

– *DA* is the MAC address of the destination.
– *SA* is the MAC address of the source.
– *Duration* is the duration of the transmission. This duration depends on the transmission mode and the destination address type (broadcast/multicast address or unicast address).

Some management frames use payload data fields to carry specific information associated to the function.

3. Beacon Frames

The beacon frame sent periodically allows to broadcast important information for accomplishing certain operations. It contains the elements shown in Table B.2 such as the value of the synchronization local clock (timestamp), the beacon-sending interval, the network identifier and the parameters needed for the station to function properly. The time unit used for the beacon intervals is of the order of a millisecond.

4. Association Frames

In order to associate, a station sends a frame containing the elements making up the request. This station in turn will receive a response to its request. The elements constituting the two types of frames (request and response) are listed in Table B.3.

The values of the status codes are summarized in Table B.10.

5. Reassociation

Similarly, to perform a reassociation, some frames to carry requests and others for responses are needed. The information relative to the two frame types are included in Table B.4.

Appendix B

Table B.3. Association frame

Order	Request	Response
1	Supported functions	Supported functions
2	Censing frequency	Status code
3	SSID	AID
4	Offered throughputs	Offered throughputs

Table B.4. Reassociation frame

Order	Request	Response
1	Supported functions	Supported functions
2	Listening frequency	Status code
3	BSSID	AID
4	SSID	Offered rates
5	Offered rates	

Table B.5. Disassociation frame

Order	Information
1	Code (reason of disassociation)

Table B.6. Probe request

Order	Information
1	SSID
2	Offered throughputs

6. Disassociation Frames

The format of this frame sent by a station wanting to terminate an existing association with a BSS or an IBSS is shown in Table B.5.

7. Probe Request Frames

A Probe request frame is sent by a station in order to collect the information needed to carry out an operation of association. The response to this request contains some elements such as the timestamp, the beacon-sending interval, the SSID, the parameters for the spread spectrum process and the available rates in the network (Table B.6).

8. Authentication Frames

The authentication frame contains the elements needed for the authentication such as the algorithm and the text proposed to the user (Table B.7).

9. Deauthentication Frames

This frame contains only the reason for the deauthentication (Table B.8).

Table B.7. Authentication frame

Order	Information
1	Deauthentication algorithm
2	Sequence number
3	Status code
4	Text to authenticate

Table B.8. Deauthentication frame

Order	Information
1	Code (reason of deauthentication)

Table B.9. TIM

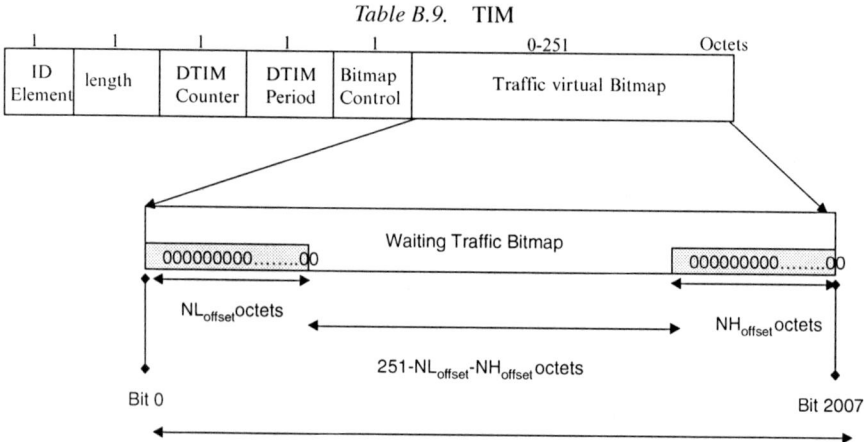

10. Traffic Indication Map Structure

Periodically, access points try to transmit data buffered for sleeping stations whose energy-conserving mechanism is activated. For those, the access point broadcasts a *traffic indication map* (TIM) structure that is formed by some information elements whose format is shown in Table B.9.

A TIM element contains four fields: *Delivery Tim* (DTIM) count, DTIM period, control and bitmap of traffic:

- The Length field indicates the size of the information field. The DTIM counter represents the number of beacons (including the current frame) to be transmitted before the next DTIM. A value of this counter equals to 0 indicates that the current TIM element is a DTIM element.

- The DTIM period field indicates the number of beacon frames between the sending of two successive DTIM frames.

- Bit 0 of the control field indicates if the frames, multicast or broadcast, are buffered in the access point. If affirmative, the bit is set to 1. The next 7 bits represent the bitmap offset, NL_{offset}, ranging between 0 and 250.

Appendix B

- The traffic virtual bitmap includes 2008 bits organized in 251 bytes. The bit i is set to 1, this implies that the access point has at least a frame of a station waiting to transmit. Bit i is equal to 0 in the case where there are no frames buffered for the station whose association identifier is equal to i. NL_{offset} indicates the number of first bytes whose bits are equal to 0 and NH_{offset} specifies the number of last bytes whose bits are equal to 0. The virtual bitmap actually sent is reduced in order not to transmit useless 0.

11. Status Codes

The different values for the status codes are summarized in Table B.10.

Table B.10. Status codes

Code	Explanation	Code	Explanation
0	Success	13	Authentication algorithm not supported
1	Failure	14	Erroneous sequence number in an authentication frame
2–9	Reserved	15	Authentication failed due to an incorrect response to the challenge
10	Cannot support the requested functions	16	Authentication failed due to time out
11	Reassociation denied as it is not possible to confirm the association	17	Association failed as the AP cannot accept new stations in
12	Association denied (unspecified reason)	18	Association failed as the station does not support all of the data rates required

19-65535 reserved

GLOSSARY

Symbols and numbers

2.5G	Cellular communication system packet switch-oriented. By this general term we refer to a system with data rates lying between those of GSM and third-generation systems. GPRS and i-mode are an example of 2.5G systems.
3G	Cellular communication system, high-speed capable and packet switch-oriented, said of third generation.
802.3	IEEE standard that specifies a carrier sense access method and a physical layer suitable for wired local area networks. It is the specification used for Ethernet.
802.11	IEEE standard that includes specifications at physical and access control layers to support wireless local area networks of IEEE 802.11 type at 1 and 2 Mbps in the 2.4 GHz band.
802.11b	Extension of the IEEE 802.11 standard to support data rates up to 5.5 and 11 Mbps in the 2.4 GHz band.
802.11a	IEEE standard employing the OFDM modulation in the 5 GHz band at a maximum speed of 54 Mbps.
802.11g	IEEE standard employing the OFDM modulation in the 2.4 GHz band at a maximum speed of 54 Mbps.
802.11e	IEEE standard that defines quality of service mechanisms for IEEE 802.11 equipment.
802.11i	IEEE working group dealing with the enhancement of the IEEE 802.11 security features.
802.11F	IEEE working group carrying out effort to normalize communication protocols between access points.
802.11c	A specification of the IEEE 802.11 family not aimed at the general public. It is just a modification of the IEEE 802.1d standard for bridging IEEE 802.11 frames (data link layer).
802.11h	Set of rules targeted to harmonize IEEE 802.11 specifications to the Hiperlan 2 standard, so that to conform to the European regulation for allowable frequencies and power levels.

802.11j — Norm to comply to the Japanese regulation, as is IEEE 802.11h for Europe.

802.15.1 — IEEE specification for the Bluetooth norm.

802.16 — Deals with technologies for the wireless local loop. There are three working groups: IEEE 802.16.1 (air interface in the 10–66 GHz band), IEEE 802.16.2 (coexistence of fixed broadband wireless access systems) and IEEE 802.16.3 (air interface in the 2–11 GHz band).

802.20 — Working group for the ≪*Mobile Broadband Wireless Access* (MBWA)≫, which specifies a standard optimized for IP transport services over wireless access interfaces in the 3.5 GHz band.

802.1X — Standard which defines a port-based access control architecture by means of several authentication procedures and mechanisms.

802.11HR — HR stands for *High Rate* (see IEEE 802.11b).

A

AAA — *Authentication, Authorization, Accounting.* Standard devoted to procedures for network access control and billing records for different telecommunication systems.

ACK — *Acknowledgement.* A response message which is sent to confirm a positive reception of a frame.

ACL — *Access Control List.* It can be constituted, for example, by a set of MAC addresses, those of stations authorized to accessing network resources.

ACL — *Asynchronous Connectionless.* A type of connection defined for the Bluetooth system, used primarily to transfer data.

ACO — *Authenticated Ciphering Offset.* Parameter used along the process of crypto key generation for a Bluetooth device.

Ad hoc — or ad hoc network. A network made of mobile units which communicate between them without an access point acting as an intermediary.

ADSL — *Asymmetric Digital Subscriber Line.* Data communication technique over typical telephone lines (128 Kbps–6 Mbps).

Glossary

AES		*Advanced Encryption Standard.* Cryptographic system based on symmetric keys. Standard developed by *(National Institute of Standards and Technology* (NIST) to replace *Data Encryption Standard* (DES) eventually.
AF		*Assured Forwarding.* Diffserv class of traffic defined for quality of service requirements.
AID		*Association Identifier.* Identifies a successful association for a given user in infrastructure mode. The duration field can include the AID. PS-Poll frames carry this identifier instead of duration.
AIFS		*Arbitration Interframe Space.* Interframe time specified in the IEEE 802.11e standard.
AP		*Access point.* Network entity equipped with a wireless interface, playing a central role in the architecture as it leads traffic to the wired part of the network (An AP is generally connected to a wired network, such as Ethernet). In infrastructure mode, the access point combines transmissions within its coverage area. Any communication traverses the AP mandatorily.
ARQ		*Automatic Repeat Request.* Correction error mechanism which retransmits corrupted frames on request issued by receiving stations.
ART		Telecommunications regulation authority in France.
ATIM		*Ad-hoc Traffic Information Map.* In ad hoc mode stations having traffic destined to other stations that are in energy-saving state, use ATIM frames to let them know. Once these beacons are heard by the intended station, the one storing the data is allowed to transmit them.

B

Back off		Random timer value selected by each station wanting to access the medium. Once it is expired (decrement is suspended during every transmission), the station is authorized to use the DIFS interframe to access the wireless channel.
Baseband		In the Bluetooth specification, the baseband layer comprises the procedures to manage the radio frequency layer.

	Beacon	Management frame broadcasted at regular intervals carrying information about the operational state of the network (synchronization, frequencies, random access periods, etc.).
	BER	*Bit Error Rate*. Ratio between the number of bits received in error and the number of bits transmitted.
	Bluetooth	Standard defined by the *Special Interest Group* (SIG) consortium for short-range systems with rates of 1 Mbps in the 2.4 GHz band. It is a wireless technology enabling voice and data communication between different electronic equipment, such as PCs, mobile phones or any peripheral in general, over short distances.
	BPSK	*Binary Phase-Shift Keying*. Two-state phase modulation (0 is encoded by a phase shift of 0°, whereas 1 is represented by a phase shift of 180°).
	BRAN	*Broadband Radio Access Network*. European project now concluded which led to the specifications for Hiperlan 1 and Hiperlan 2.
	BS	*Base Station*. A set of transmitters and receivers (transceiver system) serving a given area or cell. This general term is widely used in mobile cellular networks.
	BSS	*Basic Service Set*. A communication cell where wireless stations can transmit/receive data and get access to available network resources. Stations within the coverage area of an access point are logically associated and form a BSS.
	BSSID	*BSS Identifier*. Network identifier of a BSS cell. It corresponds to the MAC address of the concerned access point.

C

	CBR	*Constant Bit Rate*.
	CC	*Central Controller*. Central element that controls the access to a shared communication medium.
	CCA	*Clear Channel Assessment*. Function used to determine the activity on the medium in an IEEE 802.11 network.
	CCITT	*International Telegraph and Telephone Consultative Committee*. An organization, which is part of ITU that recommends telecommunications standards.

CCK	*Complementary Code Keying.* Modulation technique offering data rates of 5.5 and 11 Mbps. It is employed in the IEEE 802.11b standard.
CCMP	*Counter Mode with CBC-MAC.* Security protocol based on the AES algorithm to provide a strong data protection. It is thought to be the successor of TKIP.
CEPT	*Conférence Européenne des Postes and Télécommunications.*
CF	*Contention Free.* Random period for accessing the medium. It corresponds to DCF mode.
CFP	*Contention-Free Period.* Random period propagated in beacon frames of a wireless network. It corresponds to PCF mode.
Challenge/Response	Authentication handshake based on a challenge message issued by a server. A client cannot interpret this challenge and reply back to the server if it does not know a certain secret. A successful exchange between server and client confirms the identity of the latter.
CHAP	*Challenge-Handshake Authentication Protocol.* Authentication protocol based on a challenge/response algorithm. CHAP permits a server to authenticate a client sharing a common secret, without transmitting the secret itself never.
Ciphertext	Encrypted data.
CMR	*Conférence Mondiale des Radiocommunications.*
Collision	Simultaneous transmission of two or more stations trying to get access to a shared medium.
CRC	*Cyclic Redundancy Check.* Redundancy information obtained by dividing polynomial functions for detecting data corruption in transmitted frames at the receiver.
Cryptoanalysis	Etude de la security des procédés cryptographiques. La cryptanalyse consiste à déchiffrer un message dont on connaît généralement le procédé de encryption, mais pas les secrets.
Cryptography	Discipline including all the principles, tools and methods for data protection, in order to protect its confidentiality, integrity and avoid misuse (ISO 8732).
CS	*Carrier Sense.*

CSMA	*Carrier Sense Multiple Access.* Random access method based on carrier sense over the medium.
CSMA/CA	*Carrier Sense Multiple Access with Collision Avoidance.* Random access method based on carrier sense and collision avoidance used in IEEE 802.11-type networks. Derived by CSMA/CD to make it suitable for the context of wireless communication.
CSMA/CD	*Carrier Sense Multiple Access with Collision Detection.* Access method used in Ethernet networks, based on carrier sense and collision detection (specified in the IEEE 802.3 standard).
CTR	*Counter Mode.* Mode ensuring data confidentiality in the CCMP protocol for IEEE 802.11 networks.
CTS	*Clear to Send.* Frame emitted by a station to request control on the channel for data it wants to send. This frame type is reserved for point-to-point traffic. Used for the CSMA/CA method with RTS/CTS.
CW	*Contention Window.* Parameter intervening in the recovery algorithm after a collision (Backoff).

D

DBPSK	*Differential Binary Phase-Shift Keying.*
DCF	*Distributed Coordination Function.* Concurrent access mode used by default to control the channel. It is present in both ad hoc and infrastructure modes.
DES	*Data Encryption Standard.* Encryption algorithm based on the cryptographic symmetric technique published by IBM. It employs a 56-bit symmetric key and operates on blocks of 64 bits.
DHx	*High Data rate packet.* Type of packet available only in ACL mode for Bluetooth. Its name comes from choosing not to use any error correction code, thus yielding a better data transfer rate.
DHCP	*Dynamic Host Configuration Protocol.* Protocol used to assign IP addresses and other network parameters dynamically.
Diffserv	Architecture approach to help handle quality of service in an IP network.
DIFS	DCF IFS. *Distributed Interframe Space.* Interframe used to start operating in DCF mode.

	DMx	*Medium Data rate packet.* Type of packet available only in ACL mode for Bluetooth. Some sort of coding is applied to this type of packet, reducing the effective transfer rate to a medium level.
	DNS	*Domain Name Service.* Network service to translate a logical or domain name of a station into an IP address.
	Downlink	Descending traffic direction, generally with reference to the client station.
	DQPSK	*Differential Quadrature Phase-Shift Keying.*
	DS	*Distribution System.* Logical entity linking wireless and wired local area networks.
	DSCP	*Differentiated Services Code Point.* A field in the IP header indicating an application's class of service.
	DSSS	*Direct Sequence Spread Spectrum.* Spreading technique used in 802.11/b/g networks.
	DTIM	*Delivery Traffic Information Map.* Frame used either in infrastructure mode to manage energy conserving or in PCF mode. PCF mode starts after sending a DTIM frame. Access points storing packets or just waiting for transmission send them after emitting a DTIM frame (in energy-saving mode).
	DV	*Data Voice packet.* Type of packet for transporting data and voice traffic together in Bluetooth systems.
E		
	EAP	*Extensible Authentication Protocol.* Security protocol defined by the IETF, supporting several authentication methods.
	EAP/MD5	*Extensible Authentication Protocol/Message Digest 5.* Authentication method supporting also password verification of the station.
	EAP-TLS	*Extensible Authentication Protocol/Transport Layer Security.* Provides mutual authentication allowing the station to identify the access point as well during the association phase. EAP-TLS makes use of a shared secret and generates an encryption key after authentication automatically. A public key infrastructure (PKI) is used.
	EAPoL	*EAP Over Line.* Extension of the EAP protocol for EAP encapsulation in Ethernet frames.

EAPoW *EAP Over Wireless.* Extension of the EAP protocol for EAP encapsulation in IEEE 802.11 frames.

EDCF *Enhanced Distributed Coordination Function.* Random distributed access method supporting quality of service.

EF *Expedite Forwarding.* Diffserv class of traffic defined for real-time data.

EIRP *Effective Isotropic Radiated Power.* Without explicit indication, EIRP represents the maximum radiated power of the antenna in the preferred direction.

ESS *Extended Service Set.* Several BSSs linked together form an ESS.

ETSI *European Telecommunications Standards Institute.*

E Family Encryption algorithms (E22, E21, E3, E1) used in the Bluetooth system.

F

FCC *Federal Communications Commission.* United States government agency for telecommunications regulation.

FCS *Frame Check Sequence.* Control field of an MAC frame used for integrity check. The FCS consists of a 32-bit CRC.

FEC *Forward Error Correction.* Error control mechanism providing automatic correction capabilities at the receiver without transmitter intervention.

FHS *Frequency Hop Synchronization.* The FHS packet permits the synchronization of the frequency hop sequence before the establishment of a piconet or when an existing piconet changes master.

FHSS *Frequency-Hopping Spread Spectrum.* Spreading technique used in the IEEE 802.11 standard.

FIFO *First In First Out.* Service policy of a queue where the first arriving is the first served.

Firewall System (or network of systems) specifically configured to control traffic flows between networks. There are several firewall types, such as packet filter or application proxy-based.

FTP *File transfer protocol.* Network protocol for file transfer.

G

GAP	*Generic Access Profile.* Defines base procedures for a Bluetooth peripheral used to configure its connections.
GFSK	*Gaussian Frequency Shift Keying.* Modulation technique frequency-based.
Gigahertz	GHz. A unit of frequency equal to one billion cycles per second.
GPRS	*General Packet Radio Service.* Data packet transmission service over radio channels. It uses packet switch technology, with a theoretical data rate of 115 Kbps, better than 9.6 Kbps of the GSM system.
GSM	*Global System for Mobile communications* (named Groupe Spécial Mobile initially). European standard for cellular digital radio systems.

H

Handoff	Automatic transfer of a communication in progress from one cell to another without interruption.
HC	*Hybrid Coordinator.* Station playing the role of central controller for all the other stations within the same cell.
HCF	*Hybrid Coordination Function.* New access method for IEEE 802.11 networks supporting quality of service.
HCI	*Host Controller Interface.* Control interface for connecting a Bluetooth device to a Bluetooth module.
HEC	*Header Error Control.* Code in use at receivers to detect and/or correct communication errors in the header.
Hertz	Hz. A unit of frequency equal to one cycle per second.
HIPERLAN	*HIgh Performance Radio Local Access Network.* Standard developed by ETSI in the framework of the BRAN European project. Hiperlan 1 operates in the 5 GHz band with a data rate of 24 Mbps. Hiperlan 2 runs in the same band, but with rates ranging from 6 to 54 Mbps.
Hold	Bluetooth mode where a device is deactivated for a certain period of time.
Hot Spot	Public or private area with a high user density (enterprises, airports, coffee shops, public parks), well delimited, suitable for WLAN deployment.
HTTP	*Hypertext Transfer Protocol.* Defines the dialogue between a web server and a client browser.

HR/DSSS *High-Rate Direct-Sequence Spread Spectrum.* See DSSS.

HV *High-quality Voice packet.* Type of packet available only in SCO mode for Bluetooth. It makes no use of CRCs.

H323 Standard describing the network protocols to be implemented in H323 multimedia devices.

I

IAPP *Inter-Access Point Protocol.* Communication protocol between access points.

IBSS *Independent BSS.* IEEE 802.11 network showing no access points. The overlapping of effective radio coverages of multiple stations forms an IBSS. These stations operate in ad hoc mode.

ICV *Integrity Check Value.* Allows to perform an integrity check of data carried by an IEEE 802.11 WEP encrypted frame.

ICMP *Internet Control Message Protocol.* Protocol used for sending control messages between systems and promptly report failures.

IEEE *Institute of Electrical and Electronic Engineers.* Professional body working for the development of communications and networking standards.

IETF *Internet Engineering Task Force.* International community working for the normalization of the Internet.

IFS *Interframe Space.* Interframe period allowing to initiate a communication (DCF, PCF) and establish priorities to access a shared medium.

Inquiry Type of message emitted by a Bluetooth device to discover other Bluetooth peripherals in its radio coverage.

IP *Internet Protocol.*

IPSec *Internet Protocol Security.* Set of protocols normalized by the IETF aiming to improve the security features of IPv4 (already present in IPv6) against attacks like eavesdropping (IP sniffing), abuse of identity (IP spoofing), prediction of packet sequences, traffic replay.

IR *Infrared.*

	IrDA	*Infrared Data Association.* Organization for the developement of specifications for infrared wireless communications.
	ISM	*Industrial, Scientific and Medical.* Frequency bands reserved for industrial, scientific and medical applications (902 MHz, 2.4 GHz and 5.8 GHz).
	ISO	*International Organization for Standardization.*
	ISP	*Internet Service Provider.*
	ITU	*International Telecommunication Union.* International organization within the United Nations system for coordinating the development of global telecom networks and services.
	IV	*Initialization Vector.* Block of bits that is required to initialize crypto operations for the WEP protocol.

K

	Keystream	Encryption key.

L

	LLC	*Logical Link Control.* The highest control layer in the IEEE 802.x specification.
	LMP	*Link Manager Protocol.* Protocol for link management in the Bluetooth system.
	L2CAP	*Logical Link and Control Adaptation Protocol.* An intermediate layer between higher and lower layers in the Bluetooth protocol stack.
	L2TP	*Layer 2 Tunneling Protocol.* Protocol for virtual private networks. L2TP allows the establishment of point to point tunnels with the following security services: authentication (CHAP, PAP), confidentiality (encryption by shared secret or public key using the RC4 algorithm with 40 or 128 bits).

M

	MAC	*Medium Access Control.* General term indicating the control layer that manages the access to a shared medium among different stations.
	MAN	*Metropolitan Area Network.*
	Megahertz	MHz. A unit of frequency equal to one million cycles per second.

	MIB	*Management Information Base.* Information base containing characteristic and operational data of a system, available for lookups or modifications by means of protocols like SNMP.
	Mobile IP	Protocol defined by the IETF for managing mobility in an IP network.
	MSDU	*Mac Service Data Unit.* The higher-level data accepted by the MAC layer for delivery on the network.
	MS	*Mobile Station.* Refers to a general wireless station like PCs, laptops, PDAs.
	MT	*Mobile Terminal.* See MS.
	MTU	*Maximum Transfer Unit.* The size of the largest packet that a network protocol can transmit, for example, 1500 bytes for Ethernet.

N

	NAK	*Negative ACK.*
	NAV	*Network Allocation Vector.* Field acting as a virtual carrier sense that indicates that a transmission is in progress. During the specified period the other stations cannot access the medium.
	NIC	*Network Interface Card.* An adapter for connecting a machine to a network.

O

	OBEX	*Object Exchange Protocol.* Communication protocol for the exchange of binary objects between devices.
	OFDM	*Orthogonal Frequency Division Multiplexing.* Multi-carrier modulation technique used in IEEE 802.11a/g and Hiperlan 2 systems.
	OSA	*Open System Authentication.* Open authentication mode defined for IEEE 802.11 networks.
	OSI	*Open System Interconnection.* A layered model, each of them devoted to specific functions, describing communications between open systems.

P

	Page Scan	A mode where a Bluetooth device listens for hearing its own identifier.
	Paging	Message transmitted to send the identifier of another device in order to establish a connection with it.

Park	A mode where a Bluetooth device (slave) is active only during short periodic time intervals.
PC	*Point Coordinator.* Coordination point managing the access to the medium by the PCF method.
PCF	*Point Coordination Function.* Optional transmission mode in the IEEE 802.11 standard. In infrastructure mode only, for controlling the reservation type. The access point authorizes in succession the different stations in its polling list to transmit.
PCI	A network card for transferring data bidirectionally between the PC and the network.
PCMCIA	*Personal Computer Memory Card International Association.* Communication interface for portable computers.
PCS	*Physical Carrier Sense.*
PDA	*Personal Digital Assistant.* Pocket computer, typically used to complement a more powerful office PC or laptop, integrating multiple functions, tools and utilities worth having while on the move.
PDU	*Protocol Data Unit.*
PER	*PDU Error Ratio.* Packet error rate. Ratio between the number of packets received in error and the number of packets transmitted.
PHB	*Per Hop Behaviour.*
PHY	*Physical layer.*
Piconet	A Bluetooth network made up of eight devices communicating simultaneously during the time in which they share the same frequency channels.
PIFS	PCF IFS. Interframe used to initiate a contention-free access period.
PIN	*Personal Identification Number.* Numerical code used to enhance control access security to a user system.
PKI	*Public Key Infrastructure.* A trust architecture for key management, providing strict procedures and services for digital certificates. The comprehensive architecture includes the certificate authority, the registration authority, systems for certificate publication/distribution.
Plaintext	Unencrypted message.
PLCP	*Physical Layer Convergence Procedure Sublayer.* Sublayer defined in the IEEE 802.11 base standard. Its purpose is to adapt to the lower sublayer that is medium-dependent (infrared, DSSS or FHSS).

	PMD	*Physical Medium-Dependent.* Defined in the IEEE 802.11 standard, has basically the role of encoding and transmit the information bits coming from the convergence layer over a specific medium.
	Poll	Interrogation request.
	PPP	*Point-to-point Protocol.* Layer 2 protocol featuring multiple services, such as authentication, retransmission, load sharing. It can carry different protocol PDUs such as IP, IPX, Appletalk over serial lines (e.g. via modem or telephone network).
	PPTP	*Point-to-Point Tunneling Protocol.* A popular protocol for virtual private networks.
	PRNG	*Pseudo Random Number Generator.*
	PS	*Power Save.* Energy-saving mode.
	PS-Poll	*Power Save Polling request frame.* In energy saving mode in an IEEE 802.11 network, when a station wants to receive data from an access point, it sends a PS-Poll request.

Q

	QoS	*Quality of Service.*
	QPSK	*Quadrature Phase-Shift Keying.* Modulation technique using four phase shifts to encode 2 bits per symbol.

R

	RADIUS	*Remote Authentication Dial-In User Service.* Protocol for authentication, authorization and configuration between network access servers.
	RC4	*Ron's Code or Rivest's Cipher.* A symmetric encryption algorithm.
	RF	*Radio Frequency.*
	RFCOMM	*Serial Cable Emulation Protocol.* Protocol for serial cable emulation in the Bluetooth system.
	Roaming	Procedure allowing a mobile user to change access point without losing connection.
	RSA	Invented by Rivest, Shamir and Adleman (RSA) in 1977, and based on asymmetric cryptography, RSA is the most popular security algorithm providing essential services such as authentication, confidentiality, integrity and digital signature.
	RSN	*Robust Security Network.*

	RTS	*Request to Send or Ready To Send*. Frame resent by a station to confirm a request of control of the channel. It is emitted as a response to a CTS frame and is reserved for point to point traffic.
	RTT	*Round Trip Time*. Time to go back and forth in TCP.
	RX	*Receive or receiver*.

S

	SAFER+	Encryption algorithm used in the Bluetooth system.
	Scatternet	Aggregation of multiple piconets.
	SCO	*Synchronous Connection-Oriented*.
	Scrambling	Technique to randomize a bit stream.
	SDP	*Service Discovery Protocol*.
	SIFS	*Short IFS*. Interframe used to initiate an access period for management frames.
	SIG	*Special Interest Group*. The association for the definition of the Bluetooth norm, comprising more than 2000 companies.
	SKA	*Shared Key Authentication*. Authentication method based on a shared key or secret.
	Slot	Basic time interval. Its duration is determined depending on the characteristic of the specific physical layer.
	Sniff	Energy-saving mode. A Bluetooth device is active only during sniffing time periods, where it listens to the channel.
	SNMP	*Simple Network Management Protocol*.
	SNR	*Signal-to-Noise Ratio*.
	SoHo	*Small office Home office*. Market segment dedicated to office and home applications.
	SSH	*Secure Shell*. Communication protocol allowing secure remote access to Unix machines (especially to run commands like rlogin, rsh and rcp). SSH alleviates the typical security weaknesses related to accessing distant Unix systems.
	SSID	*Service Set Identifier*. Identifier of a wireless network.
	SSL	*Secure Socket Layer*. Secure communication protocol providing security services based on symmetric (DES, 3-DES, RCx) and asymmetric (RSA) encryption methods.
	STA	*Station*.

T

TBTT	*Target Beacon Transmission Time.*
TC	*Traffic Class.* Class of traffic defined in IEEE 802.11e.
TCP	*Transport Control Protocol.* Connection-oriented protocol for reliable data exchanges between two systems (OSI layer 4) connected by one or more IP networks.
TCP/IP	*Transmission Control Protocol/Internet Protocol.*
TDD	*Time Division Duplex.* Duplexing technique based on time frames.
TKIP	*Temporal Key Integrity Protocol.* Protocol for strong data encryption in IEEE 802.11 networks.
TIM	*Traffic Information Map.* Frame used in infrastructure mode to manage energy-saving and for PCF mode.
TPC	*Transmit Power Control.* Procedure for controlling transmission power.
TSF	*Timing Synchronization Function.* Data structure broadcasted in beacon frames in IEEE 802.11 networks to ensure synchronization of mobile stations.
TX	*Transmit or Transmitter.*
TXOP	*Transmission Opportunity.* Indicates the time interval in which a station has the right to transmit. A starting time and a maximum duration of transmissions are defined.

U

UDP	*User Datagram Protocol.* Connectionless protocol that can be used over IP networks.
UMTS	*Universal Mobile Telecommunication Service.*
Uplink	Ascending traffic direction, generally with reference to the client station.
USB	*Universal Serial Bus.* Serial bus standard to interface up to 127 peripheral devices.

V

VBR	*Variable Bit Rate.*
VCS	*Virtual Carrier Sense.* A carrier sense method based on using the NAV field of IEEE 802.11 frames.
VPN	*Virtual Private Network.*

W

	WAN	*Wide Area Network.*
	WAP	*Wireless Application Protocol.* A protocol developed to allow efficient access to Internet or web content to mobile terminals like portable phones.
	WECA	*Wireless Ethernet Compatibility Alliance.* Organization committed to certifying interoperability of IEEE 802.11b equipment. Owner of the Wi-Fi logo for IEEE 802.11b products.
	WEP	*Wired Equivalent Privacy.* Data encryption protocol defined in the IEEE 802.11 standard.
	Whitening	Technique to equalize the spectrum of a signal, making it similar to the white noise spectrum.
	Wi-Fi	*Wireless Fidelity.* Label assigned by WECA (Wi-Fi Alliance) to 802.11b-certified equipment. Name generally used to denote the IEEE 802.11b norm.
	Wi-Fi Alliance	The new name for WECA.
	WISP	*Wireless ISP.*
	WLAN	*Wireless Local Area Network.* Technology for connecting fixed or mobile stations in a limited area wirelessly.
	WLANA	*Wireless LAN Alliance.*
	WLL	Wireless Local Loop.
	WMAN	*Wireless MAN.*
	WPA	*Wi-Fi Protected Access.* Label assigned to Wi-Fi equipment by the Wi-Fi alliance, implementing stronger security mechanisms than WEP, such as TKIP.
	WPAN	*Wireless Personal Area Network.* Proximity network of small extension. Bluetooth is an example.
	WSN	*Wireless Sensor Network.*

References

ABOBA B., SIMON D., "PPP EAP/TLS Authentication Protocol", RFC 2716, IETF, October 1999.

ABOBA B., *The Unofficial 802.11 Security Wej Page*, http://www.drizzle.com/~aboba/IEEE

ARBAUGH W.A., SHANKAR N., WANG J., "Your 802.11 Network has no Clothes", *Proceedings of the First IEEE International Conference on Wireless LANs and Home Networks*, http://www.cs.umd.edu/~waa/wireless.pdf, December 2001.

BARRY J.R., *Wireless Infrared Communications*, Kluwer Academic, Boston, MA, 1994.

BLAKE S.D., BLACK M., CARLSON M., DAVIES E., WANG Z., WEISS W., "An Architecture for Differentiated Services"», *RFC 2475*, December 1998.

BLUNK L., VOLLBRECHT J., *PPP Extensible Authentication Protocol (EAP)*, RFC 2284, IETF, March 1998.

BORISOV N., GOLDBERG I., WAGNER D., "Intercepting Mobile Communications: The Insecurity of 802.11", *Proceedings of the 7th ACM International Conference on Mobile Computing and Networking*, Rome, Italy, July 2001.

BRAY J., STURMAN C.F, *Bluetooth, Connect Without Cables*, Prentice-Hall, Englewood Cliffs, NJ, 2001.

CALHOUN P. *et al.*, "Diameter Base Protocol", RFC 3588, IETF, September 2003.

CASOLE M., «WLAN Security – Status, Problems and Perspective», *Proceedings of the European Wireless 2002*, Florence, Italy, February 2002.

CHOI S., "Can EDCF Support QoS?" *IEEE 802.11-01/413*, 2001.

CHOI S. *et al.*, "Multiple Frame Exchanges during EDCF TXOP", *IEEE 802.11-02/566*, November 2002.

IEEE Standard 802.3, "CSMA/CD Access Method and Physical Layer Specifications", IEEE Project 802, Local Area Network Standards, IEEE, 1983.

L.M.S.C of the IEEE Computer Society, "Wireless LAN Medium Access Control (MAC) and Physical Layer (PHY) Specifications", *IEEE Standard 802.11*, 1999 editions, 1999.

L.M.S.C of the IEEE Computer Society, "Wireless LAN Medium Access Control (MAC) and Physical Layer (PHY) Specifications: Higher-Speed Physical Layer Extension in the 2.4 GHz Band", *IEEE Standard 802.11b*, 1999 editions, 1999.

IEEE standard 802.11a, *High-Speed Physical Layer in the 5GHz Band*, 1999 edition (supplement to 802.11-1999), 1999.

Specification of the Bluetooth System, volumes 1 and 2, version 1.0B, (spécification disponible sur le site web – http://www.bluetooth.org/spec/), December 1999.

Hiperlan Type 2: Physical Layer, ETSI 101 475, February 2001.

Bluetooth SIG, *Specification of the Bluetooth System*, version 1.1 (février 2001), version 1.2 (novembre 2003), (spécification disponible sur le site web – http://www.bluetooth.org/spec/), 2001.

Advanced Encryption Standard (AES), Federal Information Processing Standard Publication 197, National Institute of Standards and Technology, 2001.

L.M.S.C of the IEEE Computer Society, "Port-Based Network Access Control", *IEEE Standard 802.1X*, June 2001.

Specification for Radio Resource Measurement, Draft Supplement to ISO/IEC 8802-11/1999(E) ANSI/IEEE Std 802.11, 1999 edition, IEEE Std 802.11k/D0, November 2002.

Bluetooth Security White Paper, Bluetooth SIG, 2002.

"IEEE standard 802.11g-2003 Part II, Wireless LAN Medium Access Control (MAC) and Physical Layer (PHY) Specifications", Amendment 4: Futher Higher Data Rate Extension in 2.4 GHz Band, 2003.

IEEE Standard P802.11F/D5, Recommended Practice for Multi-Vendor Access Point Interoperability via Inter-Access Point Protocol Across Distribution Systems Supporting IEEE 802.11 Operation, unapproved draft, January 2003.

IEEE 802.11 WG, draft Supplement to Standard for Telecommunications and Information Exchange between Systems – LAN/MAC Specific Requirements – Part 11, *Wireless Medium Access Control (MAC) and Physical Layer (PHY) Specifications: Medium Access Control (MAC) Enhancements for Quality of Service (QoS)*, IEEE 802.11e/D4.0, 2003.

IEEE draft Supplement to Standard for Telecommunications and Information Exchange Between Systems – LAN/MAC Specific Requirements – Part 11: Wireless Medium Access Control (MAC) and Physical Layer (PHY) Specifications, *Specification for Enhanced Security*, Standard 802.11i, D3.0, 2003.

CROW B.P., WIDJAJA I., KIM J.G., SAKAI P.T., "IEEE 802.11 Wireless Local Area Networks", *IEEE Communications Magazine*, September 1997.

DOBBERTIN H., "The Status of MD5 After a Recent Attack", *RSA Laboratories' CryptoBytes*, volume 2, No. 2, 1999, ftp://ftp.rsasecurity.com/pub/cryptobytes/crypto2n2.pdf

FLUHRER S., MARTIN I., SHAMIR A., "Weaknesses in the Key Scheduling Algorithm of RC4", *Proceedings of the 8th Annual Workshop on Selected Areas in Cryptography*, August 2001.

GEIER J., *Wireless LANs*, Wiley, New York, 2000.

GEIER J., *Wireless LANs: Implementing High Performance IEEE 802.11 Networks*, 2nd edition, SAMS, 2002.

HAARSTEN J.C., "The Bluetooth Radio System", *IEEE Personal Communications Magazine*, volume 7, pp.28–36, February 2000.

HAGER C.T., MIDKIFF S.F, "An Analysis of Bluetooth Security Vulnerabilities", *IEEE Wireless Communications and Networking*, volume 3, WCNC, New Orleans, LA, pp. 16–20, 2003.

HECKER A., LABIOD H., SERHROUCHNI A., *Authentis: Through Incremental Authentication Models to Secure Interconnected Wi-Fi WLANS*, IEEE ASWN, Paris, 2002.

HOUSLEY R., WHITING D., *Temporal Key Hash*, 2001, http://grouper.ieee.org/groups/802/11/Documents/DocumentHolder/1-550.zip, 2001.

HOUSLEY R., WHITING D., FERGUSON N., *Counter with CBC-MAC*, http://csrc.nist.gov/encryption/modes/proposedmodes/ccm/ccm.pdf, 2003.

KACO M., *Sécurité des Réseaux*, Macmillan Technical Publishing, France, 1999.

LAGRANGE X., GODLEWSKI P., TABBANE S., *Réseaux GSM*, cinquième édition revue et augmentée, Hermès, Paris, 2000.

LEE W.C., *Mobile Cellular Telecommunications: Analog and Digital Systems*, 2nd edition, 1995.

LOSHIN P., *Big Book of IPsec RFCs: Internet Security Architecture*, Academic Press, New York, 2000.

MEDAHI A., AFIFI H., ZEGHLACHE D., "A New Model for Wireless VoIP speech quality Evaluation", *PIMRC 2003*, China, 2003.

MENMZES A.J., VAN OOrschot P.C., VANSTONE S., *Handbook of Applied Cryptography*, http://www.cacr.math.uwaterloo.ca/hac/, 1996.

MILLER A., BISDIKIAN C., *Bluetooth Revealed*, Prentice-Hall, Englewood Cliffs, NJ, 2001.

MANGOLD S. *et al.*, "IEEE 802.11e Wireless LAN for Quality of Service", *Proceedings of the European Wireless 2002*, Italy, 2002.

MULLER T., "Bluetooth Security Architecture version 1.0", http://www.bluetooth.com/developer/whitepaper/whitepaper.asp, 2001.

NEE R.V., *OFDM Wireless Multimedia Communications*, Artech House, Boston, MA, 2000.

O'HARA B., PETRICK A., *The IEEE 802.11 Handbook: A Designer's Companion*, IEEE Press, New York, 2001.

PARKS G., "Unfairness in 802.11 DCF Networks", *IEEE 802.11-01/052*, 2001.

PEI Z., WEIDONG L., JING W., YOUZHEN W., *Bluetooth – The Fastest Developing Wireless Technology*, volume 2, Communication Technology Proceedings, 2000, WCC – ICCT 2000.

PROAKIS J.G., *Digital Communications*, 3rd edition, McGraw-Hill, New York, 1995.

RAINA K., HARSCH A., *Commerce Security, A Beginner's Guide*, McGraw-Hill, New York, 2002.

RESCORLA E., *SSL and TLS: Designing and Building Secure Systems*, Addison-Wesely, Reading, MA, 2002.

RIGNEY C., WILLENS S., RUBENS A., SIMPSON W., *Remote Authentication Dial-In User Service (RADIUS)*, RFC 2865, IETF, June 2000.

SANTAMARIA A., LOPEZ HERNANDEZ F.J., *Wireless LAN Systems*, Artech House, Boston, MA, 1993.

SOCOLOFSKY T., KALE C., *A TCP/IP Tutorial*, RFC 1180, IETF, January 1991.

STEVENS W.R., "TCP/IP Illustrated", Addison-Wesley, Reading, MA, 1995.

TERRY J., *ODFM Wireless LANs: A Theoretical and Practical Guide*, SMAS, 2002.

WALKER J., "Unsafe at any Key Size: An Analysis of the WEP Encapsulation", *IEEE Document 802.11-00/362*, October 2000.

WALKE B., *Mobile Radio Networks: Networking, Protocols and Traffic Performance*, 2nd edition, Wiley, Chichester, UK, 2002.

WILLATS W., RIGNEY C., CALHOUN P., *RADIUS Extensions*, RFC 2869, IETF, June 2000.

WENIG(R.P., *Wireless LANs*, Academic Press, New York, 1997.

XU S., "Advances in WLAN QoS for 802.11: An Overview", Invited paper, 14th IEEE International Symposium on Personal Indoor and Mobile Radio Communication, Beiging, China, September 2003.

XU S., "Enhancement on Distributed Admission Control", *IEEE 802.11-02/745*, November 2002.

Crossbow Getting Started Guide (Rev.A, April 2005, Document 7430-0022-06) – http://www.xbow.com/Support/manuals.htm.

Crossbow MPR/MIB User Manual (Rev.B, April 2005, Document 7430-0021-06) – http://www.xbow.com/Support/manuals.htm

Crossbow MTS/MDA User Manual (Rev.B, April 2005, Document 7430-0020-03) – http://www.xbow.com/Support/manuals.htm

TinyOS Website, Documentation and Tutorial – http://www.tinyos.net

ZHAO J., GOVINDAN R., Understanding packet delivery performance in dense wireless sensor networks, SenSys 2003.

Crossbow Technical Paper, Avoiding RF interference between Wi-Fi and ZigBee – http://www.xbow.com/Products/Product_pdf_files/Wireless_pdf/ZigBeeandWi-FiInterference.pdf

POLASTRE J., BUONADONNA P., *et al.*, TinyOS Radio Stacks (ppt presentation), http://webs.cs.berkeley.edu, 2003.

POLASTRE J., HILL J., Culler D., Versatile Low Power Media Access for Wireless Sensor Networks (B-MAC), SenSys 2004.

POLASTRE J., SZEWCZYK R., CULLER D., Telos: Enabling Ultra-Low Power Wireless Research, http://www.polastre.com/papers/spots05-telos.pdf, 2005.

GAY D., LEVIS P., VON BEHREN R., WELSH M., BREWER E., and CULLER D., The nesC Language: A Holistic Approach to Networked Embedded Systems, *Proceedings of Programming Language Design and Implementation (PLDI) 2003*, June 2003.

LEVIS P., TinyOS Programming, February 2006 (available at http://www.tinyos.net).

RedMAX AN-100U Base Station User Manual (Redline ©, Doc.Ref. 70-00058-01-00-RedMAX_AN-100U_UserMan-20060515a).

RedMAX SU-O (Subscriber Unit - Outdoors) User Manual (Redline ©, Doc.Ref. 70-00057-01-00-RedMAX_SU-O_UserMan-20060516a).

RedMAX AN-100U Base Station Installation Guidelines (Redline ©, Doc.Ref. 70-00059-01-00-RedMAX_AN-100U_Installation_Guide-20060515a).

IEEE 802.15.4 2003. Part 15.4: Wireless Medium Access Control (MAC) and Physical Layer (PHY) Specifications for Low-Rate Wireless Personal Area Networks (LR-WPANs).

ZigBee Document 053474r06, Version 1.0 December 2004 Sponsored by: ZigBee Alliance.

IEEE 802.15 WPAN Group TG3c, SG4c, TG4d, Task Group 3c - mmWave, Study Group 4d and TG5 SGmban WNG.

Websites

Chapter 2

http://www.etsi.org
http://standards.ieee.org
http://www.hiperlan2.com
http://www.bluetooth.com
http://www.iec.org
http://www.cisco.com
http://www.wi-fi.org
http://www.afnet.fr
http://www.wireless-fr.org
http://www.paris-sansfil.net
http://www.orange-wifi.com
http://www.01net.com
http://seattlewireless.net
http://www.weca.net

http://www.ietf.org/html.charters/mobileip-charter.html
http://www.wi-fizone.org
http://www.wirelessbroadbandalliance.com
http://www.wimaxforum.org
http://www.telecoms.com/planetwireless
http://www.802wirelessworld.com
http://www.wi-fi.org

Chapter 3

http://www.bluetooth.org/
http://www.bluetooth.com/developer/specification/
http://www.bluetooth.com/developer/whitepaper/
http://bluetooth.ericsson.se
http://www.digianswer.com
http://www.palowireless.com
http://www.nokia.ch/french/technology/technocorner/bluetooth_einleitung.html
http://www.hardware.fr/html/articles/lire.php3?article=326&page=1
http://www.ericsson.com/bluetooth/companyove/history-bl/
http://bluez.sourceforge.net/howto/   (Bluetooth for Linux)

Chapter 5

http://ieee802.org/16/

Chapter 6

Wi-Fi

http://www.drizzle.com/~aboba/IEEE/
http://www.sss-mag.com/pdf/wireless.pdf
http://grouper.ieee.org/groups/802/11/Documents/DocumentHolder/0-362.zip
http://www.practicallynetworked.com/tools/wireless_acticles_security.htmhttp:
   //www.isaac.cs.berkeley.edu/isaac/wep-faq.html
http://www.cs.umd.edu/~waa/attack/
http://www.sublimation.org/security/localarchive/802.11/wep_attack.pdf
http://csrc.nist.gov/encryption/aes/
http://www.securiteinfo.com/crypto/802_11.shtml

Bluetooth

http://www.bluetooth.com/developer/whitepaper/
http://www.bell-labs.com
http://www.intel.com
http://www.mcommercetimes.com/technology/41/

Printed in the United States
114401LV00002B/209/A